WHAT IS LIFE?

with

MIND AND MATTER

&

AUTOBIOGRAPHICAL SKETCHES

Canto is an imprint offering a range of titles, classic and more recent, across a broad spectrum of subject areas and interests. History, literature, biography, archaeology, politics, religion, psychology, philosophy and science are all represented in Canto's specially selected list of titles, which now offers some of the best and most accessible of Cambridge publishing to a wider readership.

WHAT IS LIFE?

The Physical Aspect of the Living Cell

with

MIND AND MATTER

&

AUTOBIOGRAPHICAL SKETCHES

ERWIN SCHRÖDINGER

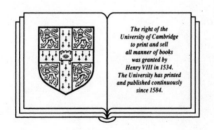

The right of the
University of Cambridge
to print and sell
all manner of books
was granted by
Henry VIII in 1534.
The University has printed
and published continuously
since 1584.

CAMBRIDGE UNIVERSITY PRESS

CAMBRIDGE
NEW YORK   PORT CHESTER
MELBOURNE   SYDNEY

Published by the Press Syndicate of the University of Cambridge
The Pitt Building, Trumpington Street, Cambridge CB2 1RP
40 West 20th Street, New York, NY 10011–4211, USA
10 Stamford Road, Oakleigh, Victoria 3166, Australia

WHAT IS LIFE?
First published 1944
Reprinted 1945, 1948, 1951, 1955, 1962

MIND AND MATTER
First published 1958
Reprinted 1959

Combined reprint 1967
Reprinted 1969, 1974, 1977, 1979, 1980, 1983,
1985, 1986, 1987, 1988, 1989 (twice)

Printed in Great Britain at the University Press, Cambridge

ISBN 0 521 42708 8 paperback

Contents

WHAT IS LIFE?

statistical meaning of entropy – Organization maintained by extracting 'order' from the environment

New laws to be expected in the organism – Reviewing the biological situation – Summarizing the physical situation – The striking contrast – Two ways of producing orderliness – The new principle is not alien to physics – The motion of a clock – Clockwork after all statistical – Nernst's Theorem – The pendulum clock is virtually at zero temperature – The relation between clockwork and organism

MIND AND MATTER

The problem – A tentative answer – Ethics

A biological blind alley? – The apparent gloom of Darwinism – Behaviour influences selection – Feigned Lamarckism – Genetic fixation of habits and skills – Dangers to intellectual evolution

WHAT IS LIFE?

THE PHYSICAL ASPECT OF THE LIVING CELL

Based on lectures delivered under the auspices of the Dublin Institute for
Advanced Studies at Trinity College, Dublin, in February 1943

To
the memory of
My Parents

Foreword

When I was a young mathematics student in the early 1950s I did not read a great deal, but what I did read – at least if I completed the book – was usually by Erwin Schrödinger. I always found his writing to be compelling, and there was an excitement of discovery, with the prospect of gaining some genuinely new understanding about this mysterious world in which we live. None of his writings possesses more of this quality than his short classic *What is Life?* – which, as I now realize, must surely rank among the most influential of scientific writings in this century. It represents a powerful attempt to comprehend some of the genuine mysteries of life, made by a physicist whose own deep insights had done so much to change the way in which we understand what the world is made of. The book's cross-disciplinary sweep was unusual for its time – yet it is written with an endearing, if perhaps disarming, modesty, at a level that makes it accessible to non-specialists and to the young who might aspire to be scientists. Indeed, many scientists who have made fundamental contributions in biology, such as J. B. S. Haldane and Francis Crick, have admitted to being strongly influenced by (although not always in complete agreement with) the broad-ranging ideas put forward here by this highly original and profoundly thoughtful physicist.

Like so many works that have had a great impact on human thinking, it makes points that, once they are grasped, have a ring of almost self-evident truth; yet they are still blindly ignored by a disconcertingly large proportion of people who should know better. How often do we still hear that quantum effects can have little relevance in the study of biology, or even that we eat food in order to gain energy? This serves to emphasize the continuing relevance that Schrödinger's *What is Life?* has for us today. It is amply worth rereading!

Roger Penrose
8 August 1991

Preface

A scientist is supposed to have a complete and thorough knowledge, at first hand, of *some* subjects and, therefore, is usually expected not to write on any topic of which he is not a master. This is regarded as a matter of *noblesse oblige*. For the present purpose I beg to renounce the *noblesse*, if any, and to be freed of the ensuing obligation. My excuse is as follows:

We have inherited from our forefathers the keen longing for unified, all-embracing knowledge. The very name given to the highest institutions of learning reminds us, that from antiquity and throughout many centuries the *universal* aspect has been the only one to be given full credit. But the spread, both in width and depth, of the multifarious branches of knowledge during the last hundred odd years has confronted us with a queer dilemma. We feel clearly that we are only now beginning to acquire reliable material for welding together the sum total of all that is known into a whole; but, on the other hand, it has become next to impossible for a single mind fully to command more than a small specialized portion of it.

I can see no other escape from this dilemma (lest our true aim be lost for ever) than that some of us should venture to embark on a synthesis of facts and theories, albeit with second-hand and incomplete knowledge of some of them – and at the risk of making fools of ourselves.

So much for my apology.

The difficulties of language are not negligible. One's native speech is a closely fitting garment, and one never feels quite at ease when it is not immediately available and has to be replaced by another. My thanks are due to Dr Inkster (Trinity College, Dublin), to Dr Padraig Browne (St Patrick's College, Maynooth) and, last but not least, to Mr S. C. Roberts. They were put to great trouble to fit the new garment on me and to even greater trouble by my occasional reluctance to give up some 'original' fashion of my own. Should some of it have survived the mitigating tendency of my friends, it is to be put at my door, not at theirs.

The head-lines of the numerous sections were originally intended to be marginal summaries, and the text of every chapter should be read *in continuo*.

E.S.

Dublin
September 1944

Homo liber nulla de re minus quam de morte cogitat; et ejus sapientia non mortis sed vitae meditatio est. SPINOZA'S *Ethics*, Pt IV, Prop. 67

(There is nothing over which a free man ponders less than death; his wisdom is, to meditate not on death but on life.)

CHAPTER I

The Classical Physicist's
Approach to the Subject

Cogito ergo sum.   DESCARTES

THE GENERAL CHARACTER AND THE PURPOSE OF THE INVESTIGATION

This little book arose from a course of public lectures, delivered by a theoretical physicist to an audience of about four hundred which did not substantially dwindle, though warned at the outset that the subject-matter was a difficult one and that the lectures could not be termed popular, even though the physicist's most dreaded weapon, mathematical deduction, would hardly be utilized. The reason for this was not that the subject was simple enough to be explained without mathematics, but rather that it was much too involved to be fully accessible to mathematics. Another feature which at least induced a semblance of popularity was the lecturer's intention to make clear the fundamental idea, which hovers between biology and physics, to both the physicist and the biologist.

For actually, in spite of the variety of topics involved, the whole enterprise is intended to convey one idea only – one small comment on a large and important question. In order not to lose our way, it may be useful to outline the plan very briefly in advance.

The large and important and very much discussed question is:

How can the events *in space and time* which take place within the spatial boundary of a living organism be accounted for by physics and chemistry?

3

The preliminary answer which this little book will endeavour to expound and establish can be summarized as follows:

The obvious inability of present-day physics and chemistry to account for such events is no reason at all for doubting that they can be accounted for by those sciences.

STATISTICAL PHYSICS. THE FUNDAMENTAL DIFFERENCE IN STRUCTURE

That would be a very trivial remark if it were meant only to stimulate the hope of achieving in the future what has not been achieved in the past. But the meaning is very much more positive, viz. that the inability, up to the present moment, is amply accounted for.

Today, thanks to the ingenious work of biologists, mainly of geneticists, during the last thirty or forty years, enough is known about the actual material structure of organisms and about their functioning to state that, and to tell precisely why, present-day physics and chemistry could not possibly account for what happens in space and time within a living organism.

The arrangements of the atoms in the most vital parts of an organism and the interplay of these arrangements differ in a fundamental way from all those arrangements of atoms which physicists and chemists have hitherto made the object of their experimental and theoretical research. Yet the difference which I have just termed fundamental is of such a kind that it might easily appear slight to anyone except a physicist who is thoroughly imbued with the knowledge that the laws of physics and chemistry are statistical throughout.[1] For it is in relation to the statistical point of view that the structure of the vital parts of living organisms differs so entirely from that of any piece of matter that we physicists and chemists have ever handled physically in our laboratories or mentally at our

---

[1] This contention may appear a little too general. The discussion must be deferred to the end of this book, pp. 82–4.

writing desks.¹ It is well-nigh unthinkable that the laws and
regularities thus discovered should happen to apply immediately to the behaviour of systems which do not exhibit the
structure on which those laws and regularities are based.

The non-physicist cannot be expected even to grasp – let
alone to appreciate the relevance of – the difference in
'statistical structure' stated in terms so abstract as I have just
used. To give the statement life and colour, let me anticipate
what will be explained in much more detail later, namely, that
the most essential part of a living cell – the chromosome fibre
– may suitably be called *an aperiodic crystal*. In physics we have
dealt hitherto only with *periodic crystals*. To a humble physicist's mind, these are very interesting and complicated
objects; they constitute one of the most fascinating and
complex material structures by which inanimate nature
puzzles his wits. Yet, compared with the aperiodic crystal,
they are rather plain and dull. The difference in structure is of
the same kind as that between an ordinary wallpaper in which
the same pattern is repeated again and again in regular
periodicity and a masterpiece of embroidery, say a Raphael
tapestry, which shows no dull repetition, but an elaborate,
coherent, meaningful design traced by the great master.

In calling the periodic crystal one of the most complex
objects of his research, I had in mind the physicist proper.
Organic chemistry, indeed, in investigating more and more
complicated molecules, has come very much nearer to that
'aperiodic crystal' which, in my opinion, is the material
carrier of life. And therefore it is small wonder that the organic
chemist has already made large and important contributions
to the problem of life, whereas the physicist has made next to
none.

¹This point of view has been emphasized in two most inspiring papers by F. G.
Donnan, *Scientia*, xxiv, no. 78 (1918), 10 ('La science physico-chimique décrit-elle
d'une façon adéquate les phénomènes biologiques?'); *Smithsonian Report for 1929*, p.
309 ('The mystery of life').

THE NAÏVE PHYSICIST'S APPROACH
TO THE SUBJECT

After having thus indicated very briefly the general idea – or rather the ultimate scope – of our investigation, let me describe the line of attack.

I propose to develop first what you might call 'a naïve physicist's ideas about organisms', that is, the ideas which might arise in the mind of a physicist who, after having learnt his physics and, more especially, the statistical foundation of his science, begins to think about organisms and about the way they behave and function and who comes to ask himself conscientiously whether he, from what he has learnt, from the point of view of his comparatively simple and clear and humble science, can make any relevant contributions to the question.

It will turn out that he can. The next step must be to compare his theoretical anticipations with the biological facts. It will then turn out that – though on the whole his ideas seem quite sensible – they need to be appreciably amended. In this way we shall gradually approach the correct view – or, to put it more modestly, the one that I propose as the correct one.

Even if I should be right in this, I do not know whether my way of approach is really the best and simplest. But, in short, it was mine. The 'naïve physicist' was myself. And I could not find any better or clearer way towards the goal than my own crooked one.

WHY ARE THE ATOMS SO SMALL?

A good method of developing 'the naïve physicist's ideas' is to start from the odd, almost ludicrous, question: Why are atoms so small? To begin with, they are very small indeed. Every little piece of matter handled in everyday life contains an enormous number of them. Many examples have been devised to bring this fact home to an audience, none of them more impressive than the one used by Lord Kelvin: Suppose that you could mark the molecules in a glass of water; then pour the contents of the glass into the ocean and stir the latter thoroughly so as to distribute the marked molecules uniformly throughout the seven seas; if then you took a glass of water anywhere out of the ocean, you would find in it about a hundred of your marked molecules.[1]

The actual sizes of atoms[2] lie between about $\frac{1}{5000}$ and $\frac{1}{2000}$ of the wave-length of yellow light. The comparison is significant, because the wave-length roughly indicates the dimensions of the smallest grain still recognizable in the microscope. Thus it will be seen that such a grain still contains thousands of millions of atoms.

Now, why are atoms so small?

Clearly, the question is an evasion. For it is not really aimed at the size of the atoms. It is concerned with the size of organisms, more particularly with the size of our own corporeal selves. Indeed, the atom is small, when referred to our civic unit of length, say the yard or the metre. In atomic physics one is accustomed to use the so-called Ångström (abbr. Å), which is the 10^{10}th part of a metre, or in decimal notation 0.0000000001 metre. Atomic diameters range between 1 and 2Å. Now those civic units (in relation to which the atoms are so small) are closely related to the size of our bodies. There is a story tracing the yard back to the humour of an English king whom his councillors asked what unit to adopt – and he stretched out his arm sideways and said: 'Take the distance from the middle of my chest to my fingertips, that will do all right.' True or not, the story is significant for our purpose. The king would naturally indicate a length comparable with that of his own body,

---

[1]You would not, of course, find exactly 100 (even if that were the exact result of the computation). You might find 88 or 95 or 107 or 112, but very improbably as few as 50 or as many as 150. A 'deviation' or 'fluctuation' is to be expected of the order of the square root of 100, i.e. 10. The statistician expresses this by stating that you would find 100 ± 10. This remark can be ignored for the moment, but will be referred to later, affording an example of the statistical $\sqrt{n}$ law.

[2]According to present-day views an atom has no sharp boundary, so that 'size' of an atom is not a very well-defined conception. But we may identify it (or, if you please, replace it) by the distance between their centres in a solid or in a liquid – not, of course, in the gaseous state, where that distance is, under normal pressure and temperature, roughly ten times as great.

knowing that anything else would be very inconvenient. With all his predilection for the Ångström unit, the physicist prefers to be told that his new suit will require six and a half yards of tweed – rather than sixty-five thousand millions of Ångströms of tweed.

It thus being settled that our question really aims at the ratio of two lengths – that of our body and that of the atom – with an incontestable priority of independent existence on the side of the atom, the question truly reads: Why must our bodies be so large compared with the atom?

I can imagine that many a keen student of physics or chemistry may have deplored the fact that every one of our sense organs, forming a more or less substantial part of our body and hence (in view of the magnitude of the said ratio) being itself composed of innumerable atoms, is much too coarse to be affected by the impact of a single atom. We cannot see or feel or hear the single atoms. Our hypotheses with regard to them differ widely from the immediate findings of our gross sense organs and cannot be put to the test of direct inspection.

Must that be so? Is there an intrinsic reason for it? Can we trace back this state of affairs to some kind of first principle, in order to ascertain and to understand why nothing else is compatible with the very laws of Nature?

Now this, for once, is a problem which the physicist is able to clear up completely. The answer to all the queries is in the affirmative.

THE WORKING OF AN ORGANISM REQUIRES EXACT PHYSICAL LAWS

If it were not so, if we were organisms so sensitive that a single atom, or even a few atoms, could make a perceptible impression on our senses – Heavens, what would life be like! To stress one point: an organism of that kind would most certainly not be capable of developing the kind of orderly thought which, after passing through a long sequence of earlier stages, ultimately results in forming, among many other ideas, the idea of an atom.

Even though we select this one point, the following considerations would essentially apply also to the functioning of organs other than the brain and the sensorial system. Nevertheless, the one and only thing of paramount interest to us in ourselves is, that we feel and think and perceive. To the physiological process which is responsible for thought and sense all the others play an auxiliary part, at least from the human point of view, if not from that of purely objective biology. Moreover, it will greatly facilitate our task to choose for investigation the process which is closely accompanied by subjective events, even though we are ignorant of the true nature of this close parallelism. Indeed, in my view, it lies outside the range of natural science and very probably of human understanding altogether.

We are thus faced with the following question: Why should an organ like our brain, with the sensorial system attached to it, of necessity consist of an enormous number of atoms, in order that its physically changing state should be in close and intimate correspondence with a highly developed thought? On what grounds is the latter task of the said organ incompatible with being, as a whole or in some of its peripheral parts which interact directly with the environment, a mechanism sufficiently refined and sensitive to respond to and register the impact of a single atom from outside?

The reason for this is, that what we call thought (1) is itself an orderly thing, and (2) can only be applied to material, i.e. to perceptions or experiences, which have a certain degree of orderliness. This has two consequences. First, a physical organization, to be in close correspondence with thought (as my brain is with my thought) must be a very well-ordered organization, and that means that the events that happen within it must obey strict physical laws, at least to a very high degree of accuracy. Secondly, the physical impressions made upon that physically well-organized system by other bodies from outside, obviously correspond to the perception and experience of the corresponding thought, forming its material, as I have called it. Therefore, the physical interactions between our system and others must, as a rule, themselves possess a certain degree of physical orderliness, that is to say, they too must obey strict physical laws to a certain degree of accuracy.

PHYSICAL LAWS REST ON ATOMIC STATISTICS AND ARE THEREFORE ONLY APPROXIMATE

And why could all this not be fulfilled in the case of an organism composed of a moderate number of atoms only and sensitive already to the impact of one or a few atoms only?

Because we know all atoms to perform all the time a completely disorderly heat motion, which, so to speak, opposes itself to their orderly behaviour and does not allow the events that happen between a small number of atoms to enrol themselves according to any recognizable laws. Only in the co-operation of an enormously large number of atoms do statistical laws begin to operate and control the behaviour of these *assemblées* with an accuracy increasing as the number of atoms involved increases. It is in that way that the events acquire truly orderly features. All the physical and chemical laws that are known to play an important part in the life of organisms are of this statistical kind; any other kind of lawfulness and orderliness that one might think of is being perpetually disturbed and made inoperative by the unceasing heat motion of the atoms.

THEIR PRECISION IS BASED ON THE LARGE NUMBER OF ATOMS INTERVENING. FIRST EXAMPLE (PARAMAGNETISM)

Let me try to illustrate this by a few examples, picked somewhat at random out of thousands, and possibly not just the best ones to appeal to a reader who is learning for the first time about this condition of things – a condition which in modern physics and chemistry is as fundamental as, say, the fact that organisms are composed of cells is in biology, or as

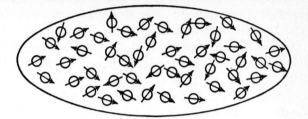

Fig. 1. Paramagnetism.

Newton's Law in astronomy, or even as the series of integers, 1, 2, 3, 4, 5, . . . in mathematics. An entire newcomer should not expect to obtain from the following few pages a full understanding and appreciation of the subject, which is associated with the illustrious names of Ludwig Boltzmann and Willard Gibbs and treated in textbooks under the name of 'statistical thermodynamics'.

If you fill an oblong quartz tube with oxygen gas and put it into a magnetic field, you find that the gas is magnetized.[1] The magnetization is due to the fact that the oxygen molecules are little magnets and tend to orientate themselves parallel to the field, like a compass needle. But you must not think that they actually all turn parallel. For if you double the field, you get double the magnetization in your oxygen body, and that proportionality goes on to extremely high field strengths, the magnetization increasing at the rate of the field you apply.

This is a particularly clear example of a purely statistical law. The orientation the field tends to produce is continually counteracted by the heat motion, which works for random orientation. The effect of this striving is, actually, only a small preference for acute over obtuse angles between the dipole axes and the field. Though the single atoms change their

---

[1] A gas is chosen, because it is simpler than a solid or a liquid; the fact that the magnetization is in this case extremely weak, will not impair the theoretical considerations.

orientation incessantly, they produce on the average (owing to their enormous number) a constant small preponderance of orientation in the direction of the field and proportional to it. This ingenious explanation is due to the French physicist P. Langevin. It can be checked in the following way. If the observed weak magnetization is really the outcome of rival tendencies, namely, the magnetic field, which aims at combing all the molecules parallel, and the heat motion, which makes for random orientation, then it ought to be possible to increase the magnetization by weakening the heat motion, that is to say, by lowering the temperature, instead of reinforcing the field. That is confirmed by experiment, which gives the magnetization inversely proportional to the absolute temperature, in quantitative agreement with theory (Curie's law). Modern equipment even enables us, by lowering the temperature, to reduce the heat motion to such insignificance that the orientating tendency of the magnetic field can assert itself, if not completely, at least sufficiently to produce a substantial fraction of 'complete magnetization'. In this case we no longer expect that double the field strength will double the magnetization, but that the latter will increase less and less with increasing field, approaching what is called 'saturation'. This expectation too is quantitatively confirmed by experiment.

Notice that this behaviour entirely depends on the large numbers of molecules which co-operate in producing the observable magnetization. Otherwise, the latter would not be constant at all, but would, by fluctuating quite irregularly from one second to the next, bear witness to the vicissitudes of the contest between heat motion and field.

SECOND EXAMPLE
(BROWNIAN MOVEMENT, DIFFUSION)

If you fill the lower part of a closed glass vessel with fog, consisting of minute droplets, you will find that the upper boundary of the fog gradually sinks, with a well-defined velocity, determined by the viscosity of the air and the size

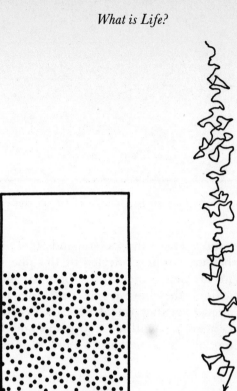

Fig. 2. Sinking fog.   Fig. 3. Brownian movement
of a sinking droplet.

and the specific gravity of the droplets. But if you look at one
of the droplets under the microscope you find that it does not
permanently sink with constant velocity, but performs a very
irregular movement, the so-called Brownian movement,
which corresponds to a regular sinking only on the average.

Now these droplets are not atoms, but they are sufficiently
small and light to be not entirely insusceptible to the impact of
one single molecule of those which hammer their surface in
perpetual impacts. They are thus knocked about and can only
on the average follow the influence of gravity.

This example shows what funny and disorderly experience
we should have if our senses were susceptible to the impact of

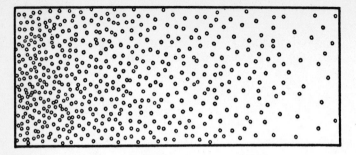

Fig. 4. Diffusion from left to right in a solution of varying concentration.

a few molecules only. There are bacteria and other organisms so small that they are strongly affected by this phenomenon. Their movements are determined by the thermic whims of the surrounding medium; they have no choice. If they had some locomotion of their own they might nevertheless succeed in getting from one place to another – but with some difficulty, since the heat motion tosses them like a small boat in a rough sea.

A phenomenon very much akin to Brownian movement is that of *diffusion*. Imagine a vessel filled with a fluid, say water, with a small amount of some coloured substance dissolved in it, say potassium permanganate, not in uniform concentration, but rather as in Fig. 4, where the dots indicate the molecules of the dissolved substance (permanganate) and the concentration diminishes from left to right. If you leave this system alone a very slow process of 'diffusion' sets in, the permanganate spreading in the direction from left to right, that is, from the places of higher concentration towards the places of lower concentration, until it is equally distributed through the water.

The remarkable thing about this rather simple and apparently not particularly interesting process is that it is in no way due, as one might think, to any tendency or force driving the permanganate molecules away from the crowded region to the less crowded one, like the population of a country spreading to those parts where there is more elbow-room. Nothing of the sort happens with our permanganate molecules. Every one of them behaves quite independently of all the others, which it very seldom meets. Every one of them, whether in a crowded region or in an empty one, suffers the same fate of being continually knocked about by the impacts of the water molecules and thereby gradually moving on in an unpredictable direction – sometimes towards the higher, sometimes towards the lower, concentrations, sometimes obliquely. The kind of motion it performs has often been compared with that of a blindfolded person on a large surface imbued with a certain desire of 'walking', but without any preference for any particular direction, and so changing his line continuously.

That this random walk of the permanganate molecules, the same for all of them, should yet produce a regular flow towards the smaller concentration and ultimately make for uniformity of distribution, is at first sight perplexing – but only at first sight. If you contemplate in Fig. 4 thin slices of approximately constant concentration, the permanganate molecules which in a given moment are contained in a particular slice will, by their random walk, it is true, be carried with equal probability to the right or to the left. But precisely in consequence of this, a plane separating two neighbouring slices will be crossed by more molecules coming from the left than in the opposite direction, simply because to the left there are more molecules engaged in random walk than there are to the right. And as long as that is so the balance will show up as a regular flow from left to right, until a uniform distribution is reached.

When these considerations are translated into mathematical language the exact law of diffusion is reached in the form of a partial differential equation

$$\frac{\partial \rho}{\partial t} = D\nabla^2 \rho,$$

which I shall not trouble the reader by explaining, though its meaning in ordinary language is again simple enough.[1] The reason for mentioning the stern 'mathematically exact' law here, is to emphasize that its physical exactitude must nevertheless be challenged in every particular application. Being based on pure chance, its validity is only approximate. If it is, as a rule, a very good approximation, that is only due to the enormous number of molecules that co-operate in the phenomenon. The smaller their number, the larger the quite haphazard deviations we must expect – and they can be observed under favourable circumstances.

THIRD EXAMPLE
(LIMITS OF ACCURACY OF MEASURING)

The last example we shall give is closely akin to the second one, but has a particular interest. A light body, suspended by a long thin fibre in equilibrium orientation, is often used by physicists to measure weak forces which deflect it from that position of equilibrium, electric, magnetic or gravitational forces being applied so as to twist it around the vertical axis. (The light body must, of course, be chosen appropriately for the particular purpose.) The continued effort to improve the accuracy of this very commonly used device of a 'torsional balance', has encountered a curious limit, most interesting in itself. In choosing lighter and lighter bodies and thinner and longer fibres – to make the balance susceptible to weaker and weaker forces – the limit was reached when the suspended body became noticeably susceptible to the impacts of the heat motion of the surrounding molecules and began to perform an incessant, irregular 'dance' about its equilibrium position, much like the trembling of the droplet in the second example. Though this behaviour sets no absolute limit to the accuracy of measurements obtained with the balance, it sets a practical one. The uncontrollable effect of the heat motion competes

[1] To wit: the concentration at any given point increases (or decreases) at a time rate proportional to the comparative surplus (or deficiency) of concentration in its infinitesimal environment. The law of heat conduction is, by the way, of exactly the same form, 'concentration' having to be replaced by 'temperature'.

with the effect of the force to be measured and makes the single deflection observed insignificant. You have to multiply observations, in order to eliminate the effect of the Brownian movement of your instrument. This example is, I think, particularly illuminating in our present investigation. For our organs of sense, after all, are a kind of instrument. We can see how useless they would be if they became too sensitive.

THE $\sqrt{n}$ RULE

So much for examples, for the present. I will merely add that there is not one law of physics or chemistry, of those that are relevant within an organism or in its interactions with its environment, that I might not choose as an example. The detailed explanation might be more complicated, but the salient point would always be the same and thus the description would become monotonous.

But I should like to add one very important quantitative statement concerning the degree of inaccuracy to be expected in any physical law, the so-called $\sqrt{n}$ law. I will first illustrate it by a simple example and then generalize it.

If I tell you that a certain gas under certain conditions of pressure and temperature has a certain density, and if I expressed this by saying that within a certain volume (of a size relevant for some experiment) there are under these conditions just n molecules of the gas, then you might be sure that if you could test my statement in a particular moment of time, you would find it inaccurate, the departure being of the order of $\sqrt{n}$. Hence if the number $n = 100$, you would find a departure of about 10, thus relative error = 10%. But if $n = 1$ million, you would be likely to find a departure of about 1,000, thus relative error = $\frac{1}{10}$%. Now, roughly speaking, this statistical law is quite general. The laws of physics and physical chemistry are inaccurate within a probable relative error of the order of $1/\sqrt{n}$, where n is the number of molecules that co-operate to bring about that law – to produce its validity within such regions of space or time (or both) that matter, for some considerations or for some particular experiment.

You see from this again that an organism must have a comparatively gross structure in order to enjoy the benefit of fairly accurate laws, both for its internal life and for its interplay with the external world. For otherwise the number of co-operating particles would be too small, the 'law' too inaccurate. The particularly exigent demand is the square root. For though a million is a reasonably large number, an accuracy of just 1 in 1,000 is not overwhelmingly good, if a thing claims the dignity of being a 'Law of Nature'.

CHAPTER 2

The Hereditary Mechanism

Das Sein ist ewig; denn Gesetze
Bewahren die lebend'gen Schätze,
Aus welchen sich das All geschmückt.[1]   GOETHE

THE CLASSICAL PHYSICIST'S EXPECTATION, FAR FROM BEING TRIVIAL, IS WRONG

Thus we have come to the conclusion that an organism and all the biologically relevant processes that it experiences must have an extremely 'many-atomic' structure and must be safeguarded against haphazard, 'single-atomic' events attaining too great importance. That, the 'naïve physicist' tells us, is essential, so that the organism may, so to speak, have sufficiently accurate physical laws on which to draw for setting up its marvellously regular and well-ordered working. How do these conclusions, reached, biologically speaking, *a priori* (that is, from the purely physical point of view), fit in with actual biological facts?

At first sight one is inclined to think that the conclusions are little more than trivial. A biologist of, say, thirty years ago might have said that, although it was quite suitable for a popular lecturer to emphasize the importance, in the organism as elsewhere, of statistical physics, the point was, in fact, rather a familiar truism. For, naturally, not only the body of an adult individual of any higher species, but every single cell composing it contains a 'cosmical' number of single atoms of

[1]Being is eternal; for laws there are to conserve the treasures of life on which the Universe draws for beauty.

every kind. And every particular physiological process that we observe, either within the cell or in its interaction with the environment, appears – or appeared thirty years ago – to involve such enormous numbers of single atoms and single atomic processes that all the relevant laws of physics and physical chemistry would be safeguarded even under the very exacting demands of statistical physics in respect of 'large numbers'; this demand I illustrated just now by the $\sqrt{n}$ rule.

Today, we know that this opinion would have been a mistake. As we shall presently see, incredibly small groups of atoms, much too small to display exact statistical laws, do play a dominating role in the very orderly and lawful events within a living organism. They have control of the observable large-scale features which the organism acquires in the course of its development, they determine important characteristics of its functioning; and in all this very sharp and very strict biological laws are displayed.

I must begin with giving a brief summary of the situation in biology, more especially in genetics – in other words, I have to summarize the present state of knowledge in a subject of which I am not a master. This cannot be helped and I apologize, particularly to any biologist, for the dilettante character of my summary. On the other hand, I beg leave to put the prevailing ideas before you more or less dogmatically. A poor theoretical physicist could not be expected to produce anything like a competent survey of the experimental evidence, which consists of a large number of long and beautifully interwoven series of breeding experiments of truly unprecedented ingenuity on the one hand and of direct observations of the living cell, conducted with all the refinement of modern microscopy, on the other.

THE HEREDITARY CODE-SCRIPT (CHROMOSOMES)

Let me use the word 'pattern' of an organism in the sense in which the biologist calls it 'the four-dimensional pattern', meaning not only the structure and functioning of that organism in the adult, or in any other particular stage, but the whole of its ontogenetic development from the fertilized egg cell to the stage of maturity, when the organism begins to reproduce itself. Now, this whole four-dimensional pattern is known to be determined by the structure of that one cell, the fertilized egg. Moreover, we know that it is essentially determined by the structure of only a small part of that cell, its nucleus. This nucleus, in the ordinary 'resting state' of the cell, usually appears as a network of chromatine,[1] distributed over the cell. But in the vitally important processes of cell division (mitosis and meiosis, see below) it is seen to consist of a set of particles, usually fibre-shaped or rod-like, called the chromosomes, which number 8 or 12 or, in man, 48. But I ought really to have written these illustrative numbers as $2 \times 4, 2 \times 6, \ldots, 2 \times 24, \ldots$, and I ought to have spoken of two sets, in order to use the expression in the customary meaning of the biologist. For though the single chromosomes are sometimes clearly distinguished and individualized by shape and size, the two sets are almost entirely alike. As we shall see in a moment, one set comes from the mother (egg cell), one from the father (fertilizing spermatozoon). It is these chromosomes, or probably only an axial skeleton fibre of what we actually see under the microscope as the chromosome, that contain in some kind of code-script the entire pattern of the individual's future development and of its functioning in the mature state. Every complete set of chromosomes contains the full code; so there are, as a rule, two copies of the latter in the fertilized egg cell, which forms the earliest stage of the future individual.

In calling the structure of the chromosome fibres a code-script we mean that the all-penetrating mind, once conceived by Laplace, to which every causal connection lay immediately open, could tell from their structure whether the egg would develop, under suitable conditions, into a black cock or into a speckled hen, into a fly or a maize plant, a rhododendron, a beetle, a mouse or a woman. To which we may add, that the appearances of the egg cells are very often remarkably similar;

---

[1] The word means 'the substance which takes on colour', viz. in a certain dyeing process used in microscopic technique.

and even when they are not, as in the case of the comparatively gigantic eggs of birds and reptiles, the difference is not
so much in the relevant structures as in the nutritive material
which in these cases is added for obvious reasons.

But the term code-script is, of course, too narrow. The
chromosome structures are at the same time instrumental in
bringing about the development they foreshadow. They are
law-code and executive power – or, to use another simile, they
are architect's plan and builder's craft – in one.

GROWTH OF THE BODY BY CELL DIVISION (MITOSIS)

How do the chromosomes behave in ontogenesis?[1]

The growth of an organism is effected by consecutive cell
divisions. Such a cell division is called mitosis. It is, in the life
of a cell, not such a very frequent event as one might expect,
considering the enormous number of cells of which our body is
composed. In the beginning the growth is rapid. The egg
divides into two 'daughter cells' which, at the next step, will
produce a generation of four, then of 8, 16, 32, 64, . . ., etc.
The frequency of division will not remain exactly the same in
all parts of the growing body, and that will break the
regularity of these numbers. But from their rapid increase we
infer by an easy computation that on the average as few as 50
or 60 successive divisions suffice to produce the number of
cells[2] in a grown man – or, say, ten times the number,[2] taking
into account the exchange of cells during lifetime. Thus, a
body cell of mine is, on the average, only the 50th or 60th
'descendant' of the egg that was I.

IN MITOSIS EVERY CHROMOSOME IS DUPLICATED

How do the chromosomes behave on mitosis? They duplicate

---

[1]Ontogenesis is the development of the individual, during its lifetime, as opposed to
phylogenesis, the development of species within geological periods.
[2]Very roughly, a hundred or a thousand (English) billions.

– both sets, both copies of the code, duplicate. The process has been intensively studied under the microscope and is of paramount interest, but much too involved to describe here in detail. The salient point is that each of the two 'daughter cells' gets a dowry of two further complete sets of chromosomes exactly similar to those of the parent cell. So all the body cells are exactly alike as regards their chromosome treasure.[1]

However little we understand the device we cannot but think that it must be in some way very relevant to the functioning of the organism, that every single cell, even a less important one, should be in possession of a complete (double) copy of the code-script. Some time ago we were told in the newspapers that in his African campaign General Montgomery made a point of having every single soldier of his army meticulously informed of all his designs. If that is true (as it conceivably might be, considering the high intelligence and reliability of his troops) it provides an excellent analogy to our case, in which the corresponding fact certainly is literally true. The most surprising fact is the doubleness of the chromosome set, maintained throughout the mitotic divisions. That it is the outstanding feature of the genetic mechanism is most strikingly revealed by the one and only departure from the rule, which we have now to discuss.

REDUCTIVE DIVISION (MEIOSIS) AND FERTILIZATION (SYNGAMY)

Very soon after the development of the individual has set in, a group of cells is reserved for producing at a later stage the so-called gametes, the sperma cells or egg cells, as the case may be, needed for the reproduction of the individual in maturity. 'Reserved' means that they do not serve other purposes in the meantime and suffer many fewer mitotic divisions. The exceptional or reductive division (called meiosis) is the one by which eventually, on maturity, the gametes are produced from these reserved cells, as a rule only a short

[1] The biologist will forgive me for disregarding in this brief summary the exceptional case of mosaics.

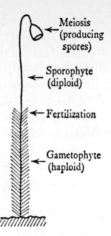

Fig. 5. Alternation of Generations.

time before syngamy is to take place. In meiosis the double chromosome set of the parent cell simply separates into two single sets, one of which goes to each of the two daughter cells, the gametes. In other words, the mitotic doubling of the number of chromosomes does not take place in meiosis, the number remains constant and thus every gamete receives only half – that is, only one complete copy of the code, not two, e.g. in man only 24, not $2 \times 24 = 48$.

Cells with only one chromosome set are called haploid (from Greek ἁπλοῦς, single). Thus the gametes are haploid, the ordinary body cells diploid (from Greek διπλοῦς, double). Individuals with three, four, . . . or generally speaking with many chromosome sets in all their body cells occur occasionally; the latter are then called triploid, tetraploid, . . ., polyploid.

In the act of syngamy the male gamete (spermatozoon) and the female gamete (egg), both haploid cells, coalesce to form the fertilized egg cell, which is thus diploid. One of its chromosome sets comes from the mother, one from the father.

HAPLOID INDIVIDUALS

One other point needs rectification. Though not indispensable for our purpose it is of real interest, since it shows that actually a fairly complete code-script of the 'pattern' is contained in every single set of chromosomes.

There are instances of meiosis not being followed shortly after by fertilization, the haploid cell (the 'gamete') undergoing meanwhile numerous mitotic cell divisions, which result in building up a complete haploid individual. This is the case in the male bee, the drone, which is produced parthenogenetically, that is, from non-fertilized and therefore haploid eggs of the queen. The drone has no father! All its body cells are haploid. If you please, you may call it a grossly exaggerated spermatozoon; and actually, as everybody knows, to function as such happens to be its one and only task in life. However, that is perhaps a ludicrous point of view. For the case is not quite unique. There are families of plants in which the haploid gamete which is produced by meiosis and is called a spore in such cases falls to the ground and, like a seed, develops into a true haploid plant comparable in size with the diploid. Fig. 5 is a rough sketch of a moss, well known in our forests. The leafy lower part is the haploid plant, called the gametophyte, because at its upper end it develops sex organs and gametes, which by mutual fertilization produce in the ordinary way the diploid plant, the bare stem with the capsule at the top. This is called the sporophyte, because it produces, by meiosis, the spores in the capsule at the top. When the capsule opens, the spores fall to the ground and develop into a leafy stem, etc. The course of events is appropriately called alternation of generations. You may, if you choose, look upon the ordinary case, man and the animals, in the same way. But the 'gametophyte' is then as a rule a very short-lived, unicellular generation, spermatozoon or egg cell as the case may be. Our body corresponds to the sporophyte. Our 'spores' are the reserved cells from which, by meiosis, the unicellular generation springs.

THE OUTSTANDING RELEVANCE OF THE
REDUCTIVE DIVISION

The important, the really fateful event in the process of reproduction of the individual is not fertilization but meiosis. One set of chromosomes is from the father, one from the mother. Neither chance nor destiny can interfere with that. Every man[1] owes just half of his inheritance to his mother, half of it to his father. That one or the other strain seems often to prevail is due to other reasons which we shall come to later. (Sex itself is, of course, the simplest instance of such prevalence.)

But when you trace the origin of your inheritance back to your grandparents, the case is different. Let me fix attention on my paternal set of chromosomes, in particular on one of them, say No. 5. It is a faithful replica either of the No. 5 my father received from his father or of the No. 5 he had received from his mother. The issue was decided by a 50:50 chance in the meiosis taking place in my father's body in November 1886 and producing the spermatozoon which a few days later was to be effective in begetting me. Exactly the same story could be repeated about chromosomes Nos. 1, 2, 3, . . ., 24 of my paternal set, and *mutatis mutandis* about every one of my maternal chromosomes. Moreover, all the 48 issues are entirely independent. Even if it were known that my paternal chromosome No. 5 came from my grandfather Josef Schrödinger, the No. 7 still stands an equal chance of being either also from him, or from his wife Marie, née Bogner.

CROSSING-OVER. LOCATION OF PROPERTIES

But pure chance has been given even a wider range in mixing the grandparental inheritance in the offspring than would appear from the preceding description, in which it has been

[1] At any rate, every *woman*. To avoid prolixity, I have excluded from this summary the highly interesting sphere of sex determination and sex-linked properties (as, for example, so-called colour blindness).

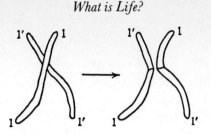

Fig. 6. *Crossing-over.* Left: the two homologous chromosomes in contact.
Right: after exchange and separation.

tacitly assumed, or even explicitly stated, that a particular
chromosome as a whole was either from the grandfather or
from the grandmother; in other words that the single chromosomes are passed on undivided. In actual fact they are not, or
not always. Before being separated in the reductive division,
say the one in the father's body, any two 'homologous'
chromosomes come into close contact with each other, during
which they sometimes exchange entire portions in the way
illustrated in Fig. 6. By this process, called 'crossing-over',
two properties situated in the respective parts of that chromosome will be separated in the grandchild, who will follow the
grandfather in one of them, the grandmother in the other one.
The act of crossing-over, being neither very rare nor very
frequent, has provided us with invaluable information regarding the location of properties in the chromosomes. For a full
account we should have to draw on conceptions not introduced before the next chapter (e.g. heterozygosy, dominance,
etc.); but as that would take us beyond the range of this little
book, let me indicate the salient point right away.

If there were no crossing-over, two properties for which the
same chromosome is responsible would always be passed on
together, no descendant receiving one of them without receiving the other as well; but two properties, due to different
chromosomes, would either stand a 50:50 chance of being
separated or they would invariably be separated – the latter
when they were situated in homologous chromosomes of the
same ancestor, which could never go together.

These rules and chances are interfered with by crossing-over. Hence the probability of this event can be ascertained by registering carefully the percentage composition of the off-spring in extended breeding experiments, suitably laid out for the purpose. In analysing the statistics, one accepts the suggestive working hypothesis that the 'linkage' between two properties situated in the same chromosome, is the less frequently broken by crossing-over, the nearer they lie to each other. For then there is less chance of the point of exchange lying between them, whereas properties located near the opposite ends of the chromosomes are separated by every crossing-over. (Much the same applies to the recombination of properties located in homologous chromosomes of the same ancestor.) In this way one may expect to get from the 'statistics of linkage' a sort of 'map of properties' within every chromosome.

These anticipations have been fully confirmed. In the cases to which tests have been thoroughly applied (mainly, but not only, *Drosophila*) the tested properties actually divide into as many separate groups, with no linkage from group to group, as there are different chromosomes (four in *Drosophila*). Within every group a linear map of properties can be drawn up which accounts quantitatively for the degree of linkage between any two out of that group, so that there is little doubt that they actually are located, and located along a line, as the rod-like shape of the chromosome suggests.

Of course, the scheme of the hereditary mechanism, as drawn up here, is still rather empty and colourless, even slightly naïve. For we have not said what exactly we understand by a property. It seems neither adequate nor possible to dissect into discrete 'properties' the pattern of an organism which is essentially a unity, a 'whole'. Now, what we actually state in any particular case is, that a pair of ancestors were different in a certain well-defined respect (say, one had blue eyes, the other brown), and that the offspring follows in this respect either one or the other. What we locate in the chromosome is the seat of this difference. (We call it, in technical language, a 'locus', or, if we think of the hypothetical material structure underlying it, a 'gene'.) Difference of property, to my view, is really the fundamental concept rather than property itself, notwithstanding the apparent linguistic and logical contradiction of this statement. The differences of properties actually are discrete, as will emerge in the next chapter when we have to speak of mutations and the dry scheme hitherto presented will, as I hope, acquire more life and colour.

MAXIMUM SIZE OF A GENE

We have just introduced the term gene for the hypothetical material carrier of a definite hereditary feature. We must now stress two points which will be highly relevant to our investigation. The first is the size – or, better, the maximum size – of such a carrier; in other words, to how small a volume can we trace the location? The second point will be the permanence of a gene, to be inferred from the durability of the hereditary pattern.

As regards the size, there are two entirely independent estimates, one resting on genetic evidence (breeding experiments), the other on cytological evidence (direct microscopic inspection). The first is, in principle, simple enough. After having, in the way described above, located in the chromosome a considerable number of different (large-scale) features (say of the *Drosophila* fly) within a particular one of its chromosomes, to get the required estimate we need only divide the measured length of that chromosome by the number of features and multiply by the cross-section. For, of course, we count as different only such features as are occasionally separated by crossing-over, so that they cannot be due to the same (microscopic or molecular) structure. On the other hand, it is clear that our estimate can only give a maximum size, because the number of features isolated by genetic analysis is continually increasing as work goes on.

The other estimate, though based on microscopic inspection, is really far less direct. Certain cells of *Drosophila* (namely, those of its salivary glands) are, for some reason, enormously enlarged, and so are their chromosomes. In them you distinguish a crowded pattern of transverse dark bands across the fibre. C. D. Darlington has remarked that the number of these bands (2,000 in the case he uses) is, though considerably larger, yet roughly of the same order of magnitude as the number of genes located in that chromosome by breeding experiments. He inclines to regard these bands as indicating the actual genes (or separations of genes). Dividing the length of the chromosome, measured in a normal-sized cell by their number (2,000), he finds the volume of a gene equal to a cube of edge 300 Å. Considering the roughness of the estimates, we may regard this to be also the size obtained by the first method.

SMALL NUMBERS

A full discussion of the bearing of statistical physics on all the facts I am recalling – or perhaps, I ought to say, of the bearing of these facts on the use of statistical physics in the living cell – will follow later. But let me draw attention at this point to the fact that 300 Å is only about 100 or 150 atomic distances in a liquid or in a solid, so that a gene contains certainly not more than about a million or a few million atoms. That number is much too small (from the $\sqrt{n}$ point of view) to entail an orderly and lawful behaviour according to statistical physics – and that means according to physics. It is too small, even if all these atoms played the same role, as they do in a gas or in a drop of liquid. And the gene is most certainly not just a homogeneous drop of liquid. It is probably a large protein molecule, in which every atom, every radical, every heterocyclic ring plays an individual role, more or less different from that played by any of the other similar atoms, radicals, or rings. This, at any rate, is the opinion of leading geneticists such as Haldane and Darlington, and we shall soon have to refer to genetic experiments which come very near to proving it.

PERMANENCE

Let us now turn to the second highly relevant question: What degree of permanence do we encounter in hereditary properties and what must we therefore attribute to the material structures which carry them?

The answer to this can really be given without any special investigation. The mere fact that we speak of hereditary properties indicates that we recognize the permanence to be almost absolute. For we must not forget that what is passed on by the parent to the child is not just this or that peculiarity, a hooked nose, short fingers, a tendency to rheumatism, haemophilia, dichromasy, etc. Such features we may conveniently select for studying the laws of heredity. But actually it is the whole (four-dimensional) pattern of the 'phenotype', the visible and manifest nature of the individual, which is reproduced without appreciable change for generations, permanent within centuries – though not within tens of thousands of years – and borne at each transmission by the material structure of the nuclei of the two cells which unite to form the fertilized egg cell. That is a marvel – than which only one is greater; one that, if intimately connected with it, yet lies on a different plane. I mean the fact that we, whose total being is entirely based on a marvellous interplay of this very kind, yet possess the power of acquiring considerable knowledge about it. I think it possible that this knowledge may advance to little short of a complete understanding – of the first marvel. The second may well be beyond human understanding.

CHAPTER 3

Mutations

Und was in schwankender Erscheinung schwebt,
Befestiget mit dauernden Gedanken.[1]   GOETHE

'JUMP-LIKE' MUTATIONS — THE WORKING-GROUND OF NATURAL SELECTION

The general facts which we have just put forward in evidence
of the durability claimed for the gene structure, are perhaps
too familiar to us to be striking or to be regarded as
convincing. Here, for once, the common saying that exceptions prove the rule is actually true. If there were no exceptions to the likeness between children and parents, we should
have been deprived not only of all those beautiful experiments
which have revealed to us the detailed mechanism of heredity,
but also of that grand, million-fold experiment of Nature,
which forges the species by natural selection and survival of
the fittest.

Let me take this last important subject as the starting-point
for presenting the relevant facts – again with an apology and a
reminder that I am not a biologist:

We know definitely, today, that Darwin was mistaken in
regarding the small, continuous, accidental variations, that
are bound to occur even in the most homogeneous population,
as the material on which natural selection works. For it has
been proved that they are not inherited. The fact is important
enough to be illustrated briefly. If you take a crop of

[1]And what in fluctuating appearance hovers,
Ye shall fix by lasting thoughts.

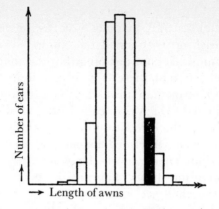

Fig. 7. Statistics of length of awns in a pure-bred crop. The black group is to be selected for sowing. (The details are not from an actual experiment, but are just set up for illustration.)

pure-strain barley, and measure, ear by ear, the length of its awns and plot the result of your statistics, you will get a bell-shaped curve as shown in Fig. 7, where the number of ears with a definite length of awn is plotted against the length. In other words: a definite medium length prevails, and deviations in either direction occur with certain frequencies. Now pick out a group of ears (as indicated by blackening) with awns noticeably beyond the average, but sufficient in number to be sown in a field by themselves and give a new crop. In making the same statistics for this, Darwin would have expected to find the corresponding curve shifted to the right. In other words, he would have expected to produce by selection an increase of the average length of the awns. That is not the case, if a truly pure-bred strain of barley has been used. The new statistical curve, obtained from the selected crop, is identical with the first one, and the same would be the case if ears with particularly short awns had been selected for seed. Selection has no effect – because the small, continuous variations are not inherited. They are obviously not based on the structure of the hereditary substance, they are accidental. But about forty years ago the Dutchman de Vries discovered that in the offspring even of thoroughly pure-bred stocks, a very small number of individuals, say two or three in tens of thousands, turn up with small but 'jump-like' changes, the expression 'jump-like' not meaning that the change is so very considerable, but that there is a discontinuity inasmuch as there are no intermediate forms between the unchanged and the few changed. De Vries called that a mutation. The significant fact is the discontinuity. It reminds a physicist of quantum theory – no intermediate energies occurring between two neighbouring energy levels. He would be inclined to call de Vries's mutation theory, figuratively, the quantum theory of biology. We shall see later that this is much more than figurative. The mutations are actually due to quantum jumps in the gene molecule. But quantum theory was but two years old when de Vries first published his discovery, in 1902. Small wonder that it took another generation to discover the intimate connection!

Mutations are inherited as perfectly as the original, unchanged characters were. To give an example, in the first crop of barley considered above a few ears might turn up with awns considerably outside the range of variability shown in Fig. 7, say with no awns at all. They might represent a de Vries mutation and would then breed perfectly true, that is to say, all their descendants would be equally awnless.

Hence a mutation is definitely a change in the hereditary treasure and has to be accounted for by some change in the hereditary substance. Actually most of the important breeding experiments, which have revealed to us the mechanism of heredity, consisted in a careful analysis of the offspring obtained by crossing, according to a preconceived plan, mutated (or, in many cases, multiply mutated) with non-mutated or with differently mutated individuals. On the other hand, by virtue of their breeding true, mutations are a suitable material on which natural selection may work and produce

Fig. 8. Heterozygous mutant. The cross marks the mutated gene.

the species as described by Darwin, by eliminating the unfit and letting the fittest survive. In Darwin's theory, you just have to substitute 'mutations' for his 'slight accidental variations' (just as quantum theory substitutes 'quantum jump' for 'continuous transfer of energy'). In all other respects little change was necessary in Darwin's theory, that is, if I am correctly interpreting the view held by the majority of biologists.[1]

LOCALIZATION. RECESSIVITY AND DOMINANCE

We must now review some other fundamental facts and notions about mutations, again in a slightly dogmatic manner, without showing directly how they spring, one by one, from experimental evidence.

We should expect a definite observed mutation to be caused by a change in a definite region in one of the chromosomes.

---

[1] Ample discussion has been given to the question, whether natural selection be aided (if not superseded) by a marked inclination of mutations to take place in a useful or favourable direction. My personal view about this is of no moment; but it is necessary to state that the eventuality of 'directed mutations' has been disregarded in all the following. Moreover, I cannot enter here on the interplay of 'switch' genes and 'polygenes', however important it be for the actual mechanism of selection and evolution.

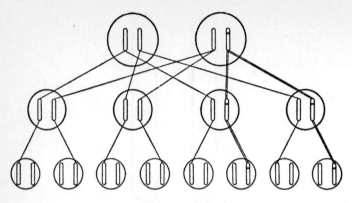

Fig. 9. Inheritance of a mutation. The straight lines across indicate the transfer of a chromosome, the double ones that of the mutated chromosome. The unaccounted-for chromosomes of the third generation come from the *mates* of the second generation, which are not included in the diagram. They are supposed to be non-relatives, free of the mutation.

And so it is. It is important to state that we know definitely that it is a change in one chromosome only, but not in the corresponding 'locus' of the homologous chromosome. Fig. 8 indicates this schematically, the cross denoting the mutated locus. The fact that only one chromosome is affected is revealed when the mutated individual (often called 'mutant') is crossed with a non-mutated one. For exactly half of the offspring exhibit the mutant character and half the normal one. That is what is to be expected as a consequence of the separation of the two chromosomes on meiosis in the mutant – as shown, very schematically, in Fig. 9. This is a 'pedigree', representing every individual (of three consecutive genera-tions) simply by the pair of chromosomes in question. Please realize that if the mutant had both its chromosomes affected, all the children would receive the same (mixed) inheritance, different from that of either parent.

But experimenting in this domain is not as simple as would appear from what has just been said. It is complicated by the second important fact, viz. that mutations are very often latent. What does that mean?

In the mutant the two 'copies of the code-script' are no

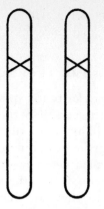

Fig. 10. Homozygous mutant, obtained in one-quarter of the descendants
either from self-fertilization of a heterozygous mutant (see Fig. 8)
or from crossing two of them.

longer identical; they present two different 'readings' or
'versions', at any rate in that one place. Perhaps it is well to
point out at once that, while it might be tempting, it would
nevertheless be entirely wrong to regard the original version
as 'orthodox', and the mutant version as 'heretic'. We have to
regard them, in principle, as being of equal right – for the
normal characters have also arisen from mutations.

What actually happens is that the 'pattern' of the individual, as a general rule, follows either the one or the other
version, which may be the normal or the mutant one. The
version which is followed is called dominant, the other
recessive; in other words, the mutation is called dominant or
recessive, according to whether it is immediately effective in
changing the pattern or not.

Recessive mutations are even more frequent than dominant
ones and are very important, though at first they do not show
up at all. To affect the pattern, they have to be present in both
chromosomes (see Fig. 10). Such individuals can be produced
when two equal recessive mutants happen to be crossed with
each other or when a mutant is crossed with itself; this is
possible in hermaphroditic plants and even happens spontaneously. An easy reflection shows that in these cases about

one-quarter of the offspring will be of this type and thus visibly exhibit the mutated pattern.

INTRODUCING SOME TECHNICAL LANGUAGE

I think it will make for clarity to explain here a few technical terms. For what I called 'version of the code-script' – be it the original one or a mutant one – the term 'allele' has been adopted. When the versions are different, as indicated in Fig. 8, the individual is called heterozygous, with respect to that locus. When they are equal, as in the non-mutated individual or in the case of Fig. 10, they are called homozygous. Thus a recessive allele influences the pattern only when homozygous, whereas a dominant allele produces the same pattern, whether homozygous or only heterozygous.

Colour is very often dominant over lack of colour (or white). Thus, for example, a pea will flower white only when it has the 'recessive allele responsible for white' in both chromosomes in question, when it is 'homozygous for white'; it will then breed true, and all its descendants will be white. But one 'red allele' (the other being white; 'heterozygous') will make it flower red, and so will two red alleles ('homozygous'). The difference of the latter two cases will only show up in the offspring, when the heterozygous red will produce some white descendants, and the homozygous red will breed true.

The fact that two individuals may be exactly alike in their outward appearance, yet differ in their inheritance, is so important that an exact differentiation is desirable. The geneticist says they have the same phenotype, but different genotype. The contents of the preceding paragraphs could thus be summarized in the brief, but highly technical statement:

A recessive allele influences the phenotype only when the genotype is homozygous.

We shall use these technical expressions occasionally, but shall recall their meaning to the reader where necessary.

THE HARMFUL EFFECT OF CLOSE-BREEDING

Recessive mutations, as long as they are only heterozygous, are of course no working-ground for natural selection. If they are detrimental, as mutations very often are, they will nevertheless not be eliminated, because they are latent. Hence quite a host of unfavourable mutations may accumulate and do no immediate damage. But they are, of course, transmitted to half of the offspring, and that has an important application to man, cattle, poultry or any other species, the good physical qualities of which are of immediate concern to us. In Fig. 9 it is assumed that a male individual (say, for concreteness, myself) carries such a recessive detrimental mutation heterozygously, so that it does not show up. Assume that my wife is free of it. Then half of our children (second line) will also carry it – again heterozygously. If all of them are again mated with non-mutated partners (omitted from the diagram, to avoid confusion), a quarter of our grandchildren, on the average, will be affected in the same way.

No danger of the evil ever becoming manifest arises, unless equally affected individuals are crossed with each other, when, as an easy reflection shows, one-quarter of their children, being homozygous, would manifest the damage. Next to self-fertilization (only possible in hermaphrodite plants) the greatest danger would be a marriage between a son and a daughter of mine. Each of them standing an even chance of being latently affected or not, one-quarter of these incestuous unions would be dangerous inasmuch as one-quarter of its children would manifest the damage. The danger factor for an incestuously bred child is thus 1:16.

In the same way the danger factor works out to be 1:64 for the offspring of a union between two ('clean-bred') grandchildren of mine who are first cousins. These do not seem to be overwhelming odds, and actually the second case is usually tolerated. But do not forget that we have analysed the consequences of only one possible latent injury in one partner of the ancestral couple ('me and my wife'). Actually both of them are quite likely to harbour more than one latent deficiency of this kind. If you know that you yourself harbour a definite one, you have to reckon with 1 out of 8 of your first cousins sharing it! Experiments with plants and animals seem to indicate that in addition to comparatively rare deficiencies of a serious kind, there seem to be a host of minor ones whose chances combine to deteriorate the offspring of close-breeding as a whole. Since we are no longer inclined to eliminate failures in the harsh way the Lacedemonians used to adopt in the Taygetos mountain, we have to take a particularly serious view about these things in the case of man, where natural selection of the fittest is largely retrenched, nay, turned to the contrary. The anti-selective effect of the modern mass slaughter of the healthy youth of all nations is hardly outweighed by the consideration that in more primitive conditions war may have had a positive value in letting the fittest tribe survive.

GENERAL AND HISTORICAL REMARKS

The fact that the recessive allele, when heterozygous, is completely overpowered by the dominant and produces no visible effect at all, is amazing. It ought at least to be mentioned that there are exceptions to this behaviour. When homozygous white snapdragon is crossed with, equally homozygous, crimson snapdragon, all the immediate descendants are intermediate in colour, i.e. they are pink (not crimson, as might be expected). A much more important case of two alleles exhibiting their influence simultaneously occurs in blood-groups – but we cannot enter into that here. I should not be astonished if at long last recessivity should turn out to be capable of degrees and to depend on the sensitivity of the tests we apply to examine the 'phenotype'.

This is perhaps the place for a word on the early history of genetics. The backbone of the theory, the law of inheritance, to successive generations, of properties in which the parents differ, and more especially the important distinction recessive-dominant, are due to the now world-famous Augustinian Abbot Gregor Mendel (1822–84). Mendel knew nothing about mutations and chromosomes. In his cloister gardens in Brünn (Brno) he made experiments on the garden pea, of which he reared different varieties, crossing them and watching their offspring in the 1st, 2nd, 3rd, . . ., generation. You might say, he experimented with mutants which he found ready-made in nature. The results he published as early as 1866 in the Proceedings of the *Naturforschender Verein in Brünn*. Nobody seems to have been particularly interested in the abbot's hobby, and nobody, certainly, had the faintest idea that his discovery would in the twentieth century become the lodestar of an entirely new branch of science, easily the most interesting of our days. His paper was forgotten and was only rediscovered in 1900, simultaneously and independently, by Correns (Berlin), de Vries (Amsterdam) and Tschermak (Vienna).

THE NECESSITY OF MUTATION BEING A RARE EVENT

So far we have tended to fix our attention on harmful mutations, which may be the more numerous; but it must be definitely stated that we do encounter advantageous mutations as well. If a spontaneous mutation is a small step in the development of the species, we get the impression that some change is 'tried out' in rather a haphazard fashion at the risk of its being injurious, in which case it is automatically eliminated. This brings out one very important point. In order to be suitable material for the work of natural selection, mutations must be rare events, as they actually are. If they were so frequent that there was a considerable chance of, say, a dozen of different mutations occurring in the same individual, the injurious ones would, as a rule, predominate over the advantageous ones and the species, instead of being improved by selection, would remain unimproved, or would perish. The comparative conservatism which results from the high degree of permanence of the genes is essential. An analogy might be sought in the working of a large manufacturing plant in a factory. For developing better methods, innovations, even if as yet unproved, must be tried out. But in order to ascertain whether the innovations improve or decrease the output, it is essential that they should be introduced one at a time, while all the other parts of the mechanism are kept constant.

We now have to review a most ingenious series of genetical research work, which will prove to be the most relevant feature of our analysis.

The percentage of mutations in the offspring, the so-called mutation rate, can be increased to a high multiple of the small natural mutation rate by irradiating the parents with X-rays or γ-rays. The mutations produced in this way differ in no way (except by being more numerous) from those occurring spontaneously, and one has the impression that every 'natural' mutation can also be induced by X-rays. In *Drosophila* many special mutations recur spontaneously again and again in the vast cultures; they have been located in the chromosome, as described on pp. 26–9, and have been given special names. There have been found even what are called 'multiple alleles', that is to say, two or more different 'versions' and 'readings' – in addition to the normal, non-mutated one – of the same place in the chromosome code; that means not only two, but three or more alternatives in that particular 'locus', any two of which are to each other in the relation 'dominant–recessive' when they occur simultaneously in their corresponding loci of the two homologous chromosomes.

The experiments on X-ray-produced mutations give the impression that every particular 'transition', say from the normal individual to a particular mutant, or conversely, has its individual 'X-ray coefficient', indicating the percentage of the offspring which turns out to have mutated in that particular way, when a unit dosage of X-ray has been applied to the parents, before the offspring was engendered.

FIRST LAW. MUTATION IS A SINGLE EVENT

Furthermore, the laws governing the induced mutation rate are extremely simple and extremely illuminating. I follow here the report of N. W. Timoféëff, in *Biological Reviews*, vol. IX, 1934. To a considerable extent it refers to that author's own beautiful work. The first law is

(1) *The increase is exactly proportional to the dosage of rays, so that one can actually speak [as I did] of a coefficient of increase.*

We are so used to simple proportionality that we are liable to underrate the far-reaching consequences of this simple law. To grasp them, we may remember that the price of a commodity, for example, is not always proportional to its amount. In ordinary times a shopkeeper may be so much impressed by your having bought six oranges from him, that, on your deciding to take after all a whole dozen, he may give it to you for less than double the price of the six. In times of scarcity the opposite may happen. In the present case, we conclude that the first half-dosage of radiation, while causing, say, one out of a thousand descendants to mutate, has not influenced the rest at all, either in the way of predisposing them for, or of immunizing them against, mutation. For otherwise the second half-dosage would not cause again just one out of a thousand to mutate. Mutation is thus not an accumulated effect, brought about by consecutive small portions of radiation reinforcing each other. It must consist in some single event occurring in one chromosome during irradiation. What kind of event?

SECOND LAW. LOCALIZATION OF THE EVENT

This is answered by the second law, viz.

(2) *If you vary the quality of the rays (wave-length) within wide limits, from soft X-rays to fairly hard γ-rays, the coefficient remains constant, provided you give the same dosage in so-called r-units*, that is to say, provided you measure the dosage by the total amount of ions produced per unit volume in a suitably chosen

standard substance during the time and at the place where the parents are exposed to the rays.

As standard substance one chooses air not only for convenience, but also for the reason that organic tissues are composed of elements of the same atomic weight as air. A lower limit for the amount of ionizations or allied processes[1] (excitations) in the tissue is obtained simply by multiplying the number of ionizations in air by the ratio of the densities. It is thus fairly obvious, and is confirmed by a more critical investigation, that the single event, causing a mutation, is just an ionization (or similar process) occurring within some 'critical' volume of the germ cell. What is the size of this critical volume? It can be estimated from the observed mutation rate by a consideration of this kind: if a dosage of 50,000 ions per cm^3 produces a chance of only $1:1000$ for any particular gamete (that finds itself in the irradiated district) to mutate in that particular way, we conclude that the critical volume, the 'target' which has to be 'hit' by an ionization for that mutation to occur, is only $\frac{1}{1000}$ of $\frac{1}{50000}$ of a cm^3, that is to say, one fifty-millionth of a cm^3. The numbers are not the right ones, but are used only by way of illustration. In the actual estimate we follow M. Delbrück, in a paper by Delbrück, N.W. Timoféëff and K.G. Zimmer,[2] which will also be the principal source of the theory to be expounded in the following two chapters. He arrives there at a size of only about ten average atomic distances cubed, containing thus only about 10^3 = a thousand atoms. The simplest interpretation of this result is that there is a fair chance of producing that mutation when an ionization (or excitation) occurs not more than about '10 atoms away' from some particular spot in the chromosome. We shall discuss this in more detail presently.

The Timoféëff report contains a practical hint which I cannot refrain from mentioning here, though it has, of course, no bearing on our present investigation. There are plenty of occasions in modern life when a human being has to be

[1] A lower limit, because these other processes escape the ionization measurement, but may be efficient in producing mutations.
[2] *Nachr. a. d. Biologie d. Ges. d. Wiss. Göttingen*, 1 (1935), 189.

exposed to X-rays. The direct dangers involved, as burns, X-ray cancer, sterilization, are well known, and protection by lead screens, lead-loaded aprons, etc., is provided, especially for nurses and doctors who have to handle the rays regularly. The point is, that even when these imminent dangers to the individual are successfully warded off, there appears to be the indirect danger of small detrimental mutations being produced in the germ cells – mutations of the kind envisaged when we spoke of the unfavourable results of close-breeding. To put it drastically, though perhaps a little naïvely, the injuriousness of a marriage between first cousins might very well be increased by the fact that their grandmother had served for a long period as an X-ray nurse. It is not a point that need worry any individual personally. But any possibility of gradually infecting the human race with unwanted latent mutations ought to be a matter of concern to the community.

CHAPTER 4

The Quantum-Mechanical Evidence

Und deines Geistes höchster Feuerflug
Hat schon am Gleichnis, hat am Bild genug.[1]   GOETHE

PERMANENCE UNEXPLAINABLE BY CLASSICAL PHYSICS

Thus, aided by the marvellously subtle instrument of X-rays (which, as the physicist remembers, revealed thirty years ago the detailed atomic lattice structures of crystals), the united efforts of biologists and physicists have of late succeeded in reducing the upper limit for the size of the microscopic structure, being responsible for a definite large-scale feature of the individual – the 'size of a gene' – and reducing it far below the estimates obtained on pp. 29–30. We are now seriously faced with the question: How can we, from the point of view of statistical physics, reconcile the facts that the gene structure seems to involve only a comparatively small number of atoms (of the order of 1,000 and possibly much less), and that nevertheless it displays a most regular and lawful activity – with a durability or permanence that borders upon the miraculous?

Let me throw the truly amazing situation into relief once again. Several members of the Habsburg dynasty have a peculiar disfigurement of the lower lip ('Habsburger Lippe'). Its inheritance has been studied carefully and published, complete with historical portraits, by the Imperial Academy of Vienna, under the auspices of the family. The feature

[1]And thy spirit's fiery flight of imagination acquiesces in an image, in a parable.

proves to be a genuinely Mendelian 'allele' to the normal form
of the lip. Fixing our attention on the portraits of a member of
the family in the sixteenth century and of his descendant,
living in the nineteenth, we may safely assume that the
material gene structure, responsible for the abnormal feature,
has been carried on from generation to generation through the
centuries, faithfully reproduced at every one of the not very
numerous cell divisions that lie between. Moreover, the
number of atoms involved in the responsible gene structure is
likely to be of the same order of magnitude as in the cases
tested by X-rays. The gene has been kept at a temperature
around 98°F during all that time. How are we to understand
that it has remained unperturbed by the disordering tendency
of the heat motion for centuries?

A physicist at the end of the last century would have been at
a loss to answer this question, if he was prepared to draw only
on those laws of Nature which he could explain and which he
really understood. Perhaps, indeed, after a short reflection on
the statistical situation he would have answered (correctly, as
we shall see): These material structures can only be molecules. Of the existence, and sometimes very high stability, of
these associations of atoms, chemistry had already acquired a
widespread knowledge at the time. But the knowledge was
purely empirical. The nature of a molecule was not understood – the strong mutual bond of the atoms which keeps a
molecule in shape was a complete conundrum to everybody.
Actually, the answer proves to be correct. But it is of limited
value as long as the enigmatic biological stability is traced
back only to an equally enigmatic chemical stability. The
evidence that two features, similar in appearance, are based
on the same principle, is always precarious as long as the
principle itself is unknown.

EXPLICABLE BY QUANTUM THEORY

In this case it is supplied by quantum theory. In the light of
present knowledge, the mechanism of heredity is closely
related to, nay, founded on, the very basis of quantum theory.

This theory was discovered by Max Planck in 1900. Modern genetics can be dated from the rediscovery of Mendel's paper by de Vries, Correns and Tschermak (1900) and from de Vries's paper on mutations (1901–3). Thus the births of the two great theories nearly coincide, and it is small wonder that both of them had to reach a certain maturity before the connection could emerge. On the side of quantum theory it took more than a quarter of a century till in 1926–7 the quantum theory of the chemical bond was outlined in its general principles by W. Heitler and F. London. The Heitler–London theory involves the most subtle and intricate conceptions of the latest development of quantum theory (called 'quantum mechanics' or 'wave mechanics'). A presentation without the use of calculus is well-nigh impossible or would at least require another little volume like this. But fortunately, now that all work has been done and has served to clarify our thinking, it seems to be possible to point out in a more direct manner the connection between 'quantum jumps' and mutations, to pick out at the moment the most conspicuous item. That is what we attempt here.

QUANTUM THEORY – DISCRETE STATES – QUANTUM JUMPS

The great revelation of quantum theory was that features of discreteness were discovered in the Book of Nature, in a context in which anything other than continuity seemed to be absurd according to the views held until then.

The first case of this kind concerned energy. A body on the large scale changes its energy continuously. A pendulum, for instance, that is set swinging is gradually slowed down by the resistance of the air. Strangely enough, it proves necessary to admit that a system of the order of the atomic scale behaves differently. On grounds upon which we cannot enter here, we have to assume that a small system can by its very nature possess only certain discrete amounts of energy, called its peculiar energy levels. The transition from one state to another is a rather mysterious event, which is usually called a 'quantum jump'.

But energy is not the only characteristic of a system. Take again our pendulum, but think of one that can perform different kinds of movement, a heavy ball suspended by a string from the ceiling. It can be made to swing in a north–south or east–west or any other direction or in a circle or in an ellipse. By gently blowing the ball with a bellows, it can be made to pass continuously from one state of motion to any other.

For small-scale systems most of these or similar characteristics – we cannot enter into details – change discontinuously. They are 'quantized', just as the energy is.

The result is that a number of atomic nuclei, including their bodyguards of electrons, when they find themselves close to each other, forming 'a system', are unable by their very nature to adopt any arbitrary configuration we might think of. Their very nature leaves them only a very numerous but discrete series of 'states' to choose from.[1] We usually call them levels or energy levels, because the energy is a very relevant part of the characteristic. But it must be understood that the complete description includes much more than just the energy. It is virtually correct to think of a state as meaning a definite configuration of all the corpuscles.

The transition from one of these configurations to another is a quantum jump. If the second one has the greater energy ('is a higher level'), the system must be supplied from outside with at least the difference of the two energies to make the transition possible. To a lower level it can change spontaneously, spending the surplus of energy in radiation.

MOLECULES

Among the discrete set of states of a given selection of atoms there need not necessarily but there may be a lowest level, implying a close approach of the nuclei to each other. Atoms

[1] I am adopting the version which is usually given in popular treatment and which suffices for our present purpose. But I have the bad conscience of one who perpetuates a convenient error. The true story is much more complicated, inasmuch as it includes the occasional indeterminateness with regard to the state the system is in.

in such a state form a molecule. The point to stress here is, that the molecule will of necessity have a certain stability; the configuration cannot change, unless at least the energy difference, necessary to 'lift' it to the next higher level, is supplied from outside. Hence this level difference, which is a well-defined quantity, determines quantitatively the degree of stability of the molecule. It will be observed how intimately this fact is linked with the very basis of quantum theory, viz. with the discreteness of the level scheme.

I must beg the reader to take it for granted that this order of ideas has been thoroughly checked by chemical facts; and that it has proved successful in explaining the basic fact of chemical valency and many details about the structure of molecules, their binding-energies, their stabilities at different temperatures, and so on. I am speaking of the Heitler–London theory, which, as I said, cannot be examined in detail here.

THEIR STABILITY DEPENDENT ON TEMPERATURE

We must content ourselves with examining the point which is of paramount interest for our biological question, namely, the stability of a molecule at different temperatures. Take our system of atoms at first to be actually in its state of lowest energy. The physicist would call it a molecule at the absolute zero of temperature. To lift it to the next higher state or level a definite supply of energy is required. The simplest way of trying to supply it is to 'heat up' your molecule. You bring it into an environment of higher temperature ('heat bath'), thus allowing other systems (atoms, molecules) to impinge upon it. Considering the entire irregularity of heat motion, there is no sharp temperature limit at which the 'lift' will be brought about with certainty and immediately. Rather, at any temperature (different from absolute zero) there is a certain smaller or greater chance for the lift to occur, the chance increasing of course with the temperature of the heat bath. The best way to express this chance is to indicate the average time you will have to wait until the lift takes place, the 'time of expectation'.

From an investigation, due to M. Polanyi and E. Wigner,[1] the 'time of expectation' largely depends on the ratio of two energies, one being just the energy difference itself that is required to effect the lift (let us write W for it), the other one characterizing the intensity of the heat motion at the temperature in question (let us write T for the absolute temperature and kT for the characteristic energy).[2] It stands to reason that the chance for effecting the lift is smaller, and hence that the time of expectation is longer, the higher the lift itself compared with the average heat energy, that is to say, the greater the ratio $W{:}kT$. What is amazing is how enormously the time of expectation depends on comparatively small changes of the ratio $W{:}kT$. To give an example (following Delbrück): for W 30 times kT the time of expectation might be as short as $\frac{1}{10}$s., but would rise to 16 months when W is 50 times kT, to 30,000 years when W is 60 times kT!

MATHEMATICAL INTERLUDE

It might be as well to point out in mathematical language – for those readers to whom it appeals – the reason for this enormous sensitivity to changes in the level step or temperature, and to add a few physical remarks of a similar kind. The reason is that the time of expectation, call it t, depends on the ratio W/kT by an exponential function, thus

$$t = \tau e^{W/kT}.$$

τ is a certain small constant of the order of 10^{-13} or 10^{-14}s. Now, this particular exponential function is not an accidental feature. It recurs again and again in the statistical theory of heat, forming, as it were, its backbone. It is a measure of the improbability of an energy amount as large as W gathering accidentally in some particular part of the system, and it is this improbability which increases so enormously when a considerable multiple of the 'average energy' kT is required.

[1] *Zeitschrift für Physik*, Chemie (A), Haber-Band (1928), p. 439.
[2] k is a numerically known constant, called Boltzmann's constant; $\frac{3}{2}kT$ is the average kinetic energy of a gas atom at temperature T.

Actually a $W = 30kT$ (see the example quoted above) is already extremely rare. That it does not yet lead to an enormously long time of expectation (only $\frac{1}{10}$s. in our example) is, of course, due to the smallness of the factor τ. This factor has a physical meaning. It is of the order of the period of the vibrations which take place in the system all the time. You could, very broadly, describe this factor as meaning that the chance of accumulating the required amount W, though very small, recurs again and again 'at every vibration', that is to say, about 10^{13} or 10^{14} times during every second.

FIRST AMENDMENT

In offering these considerations as a theory of the stability of the molecule it has been tacitly assumed that the quantum jump which we called the 'lift' leads, if not to a complete disintegration, at least to an essentially different configuration of the same atoms – an isomeric molecule, as the chemist would say, that is, a molecule composed of the same atoms in a different arrangement (in the application to biology it is going to represent a different 'allele' in the same 'locus' and the quantum jump will represent a mutation).

To allow of this interpretation two points must be amended in our story, which I purposely simplified to make it at all intelligible. From the way I told it, it might be imagined that only in its very lowest state does our group of atoms form what we call a molecule and that already the next higher state is 'something else'. That is not so. Actually the lowest level is followed by a crowded series of levels which do not involve any appreciable change in the configuration as a whole, but only correspond to those small vibrations among the atoms which we have mentioned above. They, too, are 'quantized', but with comparatively small steps from one level to the next. Hence the impacts of the particles of the 'heat bath' may suffice to set them up already at fairly low temperature. If the molecule is an extended structure, you may conceive these vibrations as high-frequency sound waves, crossing the molecule without doing it any harm.

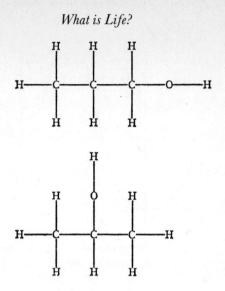

Fig. 11. The two isomers of propyl-alcohol.

So the first amendment is not very serious: we have to disregard the 'vibrational fine-structure' of the level scheme. The term 'next higher level' has to be understood as meaning the next level that corresponds to a relevant change of configuration.

SECOND AMENDMENT

The second amendment is far more difficult to explain, because it is concerned with certain vital, but rather complicated, features of the scheme of relevantly different levels. The free passage between two of them may be obstructed, quite apart from the required energy supply; in fact, it may be obstructed even from the higher to the lower state.

Let us start from the empirical facts. It is known to the chemist that the same group of atoms can unite in more than one way to form a molecule. Such molecules are called isomeric ('consisting of the same parts'; ἴσος = same, μέρος = part). Isomerism is not an exception, it is the rule. The larger

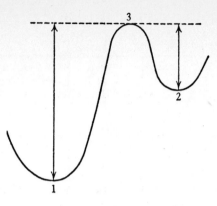

Fig. 12. Energy threshold (3) between the isomeric levels (1) and (2).
The arrows indicate the minimum energies required for transition.

the molecule, the more isomeric alternatives are offered. Fig.
11 shows one of the simplest cases, the two kinds of propyl-
alcohol, both consisting of 3 carbons (C), 8 hydrogens (H),
1 oxygen (O).[1] The latter can be interposed between any
hydrogen and its carbon, but only the two cases shown in our
figure are different substances. And they really are. All their
physical and chemical constants are distinctly different. Also
their energies are different, they represent 'different levels'.

The remarkable fact is that both molecules are perfectly
stable, both behave as though they were 'lowest states'. There
are no spontaneous transitions from either state towards the
other.

The reason is that the two configurations are not neighbouring configurations. The transition from one to the other
can only take place over intermediate configurations which
have a greater energy than either of them. To put it crudely,
the oxygen has to be extracted from one position and has to be
inserted into the other. There does not seem to be a way of
doing that without passing through configurations of considerably higher energy. The state of affairs is sometimes

[1]Models, in which C, H and O were represented by black, white and red wooden balls
respectively, were exhibited at the lecture. I have not reproduced them here, because
their likeness to the actual molecules is not appreciably greater than that of Fig. 11.

figuratively pictured as in Fig. 12, in which 1 and 2 represent the two isomers, 3 the 'threshold' between them, and the two arrows indicate the 'lifts', that is to say, the energy supplies required to produce the transition from state 1 to state 2 or from state 2 to state 1, respectively.

Now we can give our 'second amendment', which is that transitions of this 'isomeric' kind are the only ones in which we shall be interested in our biological application. It was these we had in mind when explaining 'stability' on pp. 49–51. The 'quantum jump' which we mean is the transition from one relatively stable molecular configuration to another. The energy supply required for the transition (the quantity denoted by W) is not the actual level difference, but the step from the initial level up to the threshold (see the arrows in Fig.12).

Transitions with no threshold interposed between the initial and the final state are entirely uninteresting, and that not only in our biological application. They have actually nothing to contribute to the chemical stability of the molecule. Why? They have no lasting effect, they remain unnoticed. For, when they occur, they are almost immediately followed by a relapse into the initial state, since nothing prevents their return.

CHAPTER 5

Delbrück's Model Discussed and Tested

Sane sicut lux seipsam et tenebras manifestat, sic veritas
norma sui et falsi est.[1]   SPINOZA, *Ethics*, Pt II, Prop. 43.

THE GENERAL PICTURE OF THE HEREDITARY SUBSTANCE

From these facts emerges a very simple answer to our
question, namely: Are these structures, composed of comparatively few atoms, capable of withstanding for long periods the
disturbing influence of heat motion to which the hereditary
substance is continually exposed? We shall assume the structure of a gene to be that of a huge molecule, capable only of
discontinuous change, which consists in a rearrangement of
the atoms and leads to an isomeric[2] molecule. The rearrangement may affect only a small region of the gene, and a vast
number of different rearrangements may be possible. The
energy thresholds, separating the actual configuration from
any possible isomeric ones, have to be high enough (compared
with the average heat energy of an atom) to make the
change-over a rare event. These rare events we shall identify
with spontaneous mutations.

The later parts of this chapter will be devoted to putting
this general picture of a gene and of mutation (due mainly to
the German physicist M. Delbrück) to the test, by comparing

---

[1] Truly, as light manifests itself and darkness, thus truth is the standard of itself and of error.
[2] For convenience I shall continue to call it an isomeric transition, though it would be absurd to exclude the possibility of any exchange with the environment.

it in detail with genetical facts. Before doing so, we may fittingly make some comment on the foundation and general nature of the theory.

THE UNIQUENESS OF THE PICTURE

Was it absolutely essential for the biological question to dig up the deepest roots and found the picture on quantum mechanics? The conjecture that a gene is a molecule is today, I dare say, a commonplace. Few biologists, whether familiar with quantum theory or not, would disagree with it. On p. 47 we ventured to put it into the mouth of a pre-quantum physicist, as the only reasonable explanation of the observed permanence. The subsequent considerations about isomerism, threshold energy, the paramount role of the ratio $W:kT$ in determining the probability of an isomeric transition – all that could very well be introduced on a purely empirical basis, at any rate without drawing on quantum theory. Why did I so strongly insist on the quantum-mechanical point of view, though I could not really make it clear in this little book and may well have bored many a reader?

Quantum mechanics is the first theoretical aspect which accounts from first principles for all kinds of aggregates of atoms actually encountered in Nature. The Heitler–London bondage is a unique, singular feature of the theory, not invented for the purpose of explaining the chemical bond. It comes in quite by itself, in a highly interesting and puzzling manner, being forced upon us by entirely different considerations. It proves to correspond exactly with the observed chemical facts, and, as I said, it is a unique feature, well enough understood to tell with reasonable certainty that 'such a thing could not happen again' in the further development of quantum theory.

Consequently, we may safely assert that there is no alternative to the molecular explanation of the hereditary substance. The physical aspect leaves no other possibility to account for its permanence. If the Delbrück picture should fail, we would have to give up further attempts. That is the first point I wish to make.

SOME TRADITIONAL MISCONCEPTIONS

But it may be asked: Are there really no other endurable structures composed of atoms except molecules? Does not a gold coin, for example, buried in a tomb for a couple of thousand years, preserve the traits of the portrait stamped on it? It is true that the coin consists of an enormous number of atoms, but surely we are in this case not inclined to attribute the mere preservation of shape to the statistics of large numbers. The same remark applies to a neatly developed batch of crystals we find embedded in a rock, where it must have been for geological periods without changing.

That leads us to the second point I want to elucidate. The cases of a molecule, a solid, a crystal are not really different. In the light of present knowledge they are virtually the same. Unfortunately, school teaching keeps up certain traditional views, which have been out of date for many years and which obscure the understanding of the actual state of affairs.

Indeed, what we have learnt at school about molecules does not give the idea that they are more closely akin to the solid state than to the liquid or gaseous state. On the contrary, we have been taught to distinguish carefully between a physical change, such as melting or evaporation in which the molecules are preserved (so that, for example, alcohol, whether solid, liquid or a gas, always consists of the same molecules, C_2H_6O), and a chemical change, as, for example, the burning of alcohol,

$$C_2H_6O + 3O_2 = 2CO_2 + 3H_2O,$$

where an alcohol molecule and three oxygen molecules undergo a rearrangement to form two molecules of carbon dioxide and three molecules of water.

About crystals, we have been taught that they form three-fold periodic lattices, in which the structure of the single molecule is sometimes recognizable, as in the case of alcohol and most organic compounds, while in other crystals, e.g. rock-salt (NaCl), NaCl molecules cannot be unequivocally

delimited, because every Na atom is symmetrically surrounded by six Cl atoms, and vice versa, so that it is largely
arbitrary what pairs, if any, are regarded as molecular
partners.

Finally, we have been told that a solid can be crystalline or
not, and in the latter case we call it amorphous.

DIFFERENT 'STATES' OF MATTER

Now I would not go so far as to say that all these statements
and distinctions are quite wrong. For practical purposes they
are sometimes useful. But in the true aspect of the structure of
matter the limits must be drawn in an entirely different way.
The fundamental distinction is between the two lines of the
following scheme of 'equations':

$$\text{molecule} = \text{solid} \quad = \text{crystal.}$$
$$\text{gas} \quad\quad = \text{liquid} = \text{amorphous.}$$

We must explain these statements briefly. The so-called
amorphous solids are either not really amorphous or not really
solid. In 'amorphous' charcoal fibre the rudimentary structure of the graphite crystal has been disclosed by X-rays. So
charcoal is a solid, but also crystalline. Where we find no
crystalline structure we have to regard the thing as a liquid
with very high 'viscosity' (internal friction). Such a substance
discloses by the absence of a well-defined melting temperature
and of a latent heat of melting that it is not a true solid. When
heated it softens gradually and eventually liquefies without
discontinuity. (I remember that at the end of the first Great
War we were given in Vienna an asphalt-like substance as a
substitute for coffee. It was so hard that one had to use a chisel
or a hatchet to break the little brick into pieces, when it would
show a smooth, shell-like cleavage. Yet, given time, it would
behave as a liquid, closely packing the lower part of a vessel in
which you were unwise enough to leave it for a couple of
days.)

The continuity of the gaseous and liquid state is a well-
known story. You can liquefy any gas without discontinuity

by taking your way 'around' the so-called critical point. But we shall not enter on this here.

THE DISTINCTION THAT REALLY MATTERS

We have thus justified everything in the above scheme, except the main point, namely, that we wish a molecule to be regarded as a solid = crystal.

The reason for this is that the atoms forming a molecule, whether there be few or many of them, are united by forces of exactly the same nature as the numerous atoms which build up a true solid, a crystal. The molecule presents the same solidity of structure as a crystal. Remember that it is precisely this solidity on which we draw to account for the permanence of the gene!

The distinction that is really important in the structure of matter is whether atoms are bound together by those 'solidifying' Heitler–London forces or whether they are not. In a solid and in a molecule they all are. In a gas of single atoms (as e.g. mercury vapour) they are not. In a gas composed of molecules, only the atoms within every molecule are linked in this way.

THE APERIODIC SOLID

A small molecule might be called 'the germ of a solid'. Starting from such a small solid germ, there seem to be two different ways of building up larger and larger associations. One is the comparatively dull way of repeating the same structure in three directions again and again. That is the way followed in a growing crystal. Once the periodicity is established, there is no definite limit to the size of the aggregate. The other way is that of building up a more and more extended aggregate without the dull device of repetition. That is the case of the more and more complicated organic molecule in which every atom, and every group of atoms, plays an individual role, not entirely equivalent to that of many others (as is the case in a periodic structure). We might quite properly call that an aperiodic crystal or solid and express our hypothesis by saying: We believe a gene – or perhaps the whole chromosome fibre[1] – to be an aperiodic solid.

It has often been asked how this tiny speck of material, the nucleus of the fertilized egg, could contain an elaborate code-script involving all the future development of the organism. A well-ordered association of atoms, endowed with sufficient resistivity to keep its order permanently, appears to be the only conceivable material structure that offers a variety of possible ('isomeric') arrangements, sufficiently large to embody a complicated system of 'determinations' within a small spatial boundary. Indeed, the number of atoms in such a structure need not be very large to produce an almost unlimited number of possible arrangements. For illustration, think of the Morse code. The two different signs of dot and dash in well-ordered groups of not more than four allow of thirty different specifications. Now, if you allowed yourself the use of a third sign, in addition to dot and dash, and used groups of not more than ten, you could form 88,572 different 'letters'; with five signs and groups up to 25, the number is 372,529,029,846,191,405.

It may be objected that the simile is deficient, because our Morse signs may have different composition (e.g. $-\,-$ and $\cdots\!-$) and thus they are a bad analogue for isomerism. To remedy this defect, let us pick, from the third example, only the combinations of exactly 25 symbols and only those containing exactly 5 out of each of the supposed 5 types (5 dots, 5 dashes, etc.). A rough count gives you the number of combinations as 62,330,000,000,000, where the zeros on the right stand for figures which I have not taken the trouble to compute.

Of course, in the actual case, by no means 'every' arrangement of the group of atoms will represent a possible molecule; moreover, it is not a question of a code to be adopted

[1]That it is highly flexible is no objection; so is a thin copper wire.

arbitrarily, for the code-script must itself be the operative factor bringing about the development. But, on the other hand, the number chosen in the example (25) is still very small, and we have envisaged only the simple arrangements in one line. What we wish to illustrate is simply that with the molecular picture of the gene it is no longer inconceivable that the miniature code should precisely correspond with a highly complicated and specified plan of development and should somehow contain the means to put it into operation.

COMPARISON WITH FACTS: DEGREE OF STABILITY; DISCONTINUITY OF MUTATIONS

Now let us at last proceed to compare the theoretical picture with the biological facts. The first question obviously is, whether it can really account for the high degree of permanence we observe. Are threshold values of the required amount – high multiples of the average heat energy kT – reasonable, are they within the range known from ordinary chemistry? That question is trivial; it can be answered in the affirmative without inspecting tables. The molecules of any substance which the chemist is able to isolate at a given temperature must at that temperature have a lifetime of at least minutes. (That is putting it mildly; as a rule they have much more.) Thus the threshold values the chemist encounters are of necessity precisely of the order of magnitude required to account for practically any degree of permanence the biologist may encounter; for we recall from p. 51 that thresholds varying within a range of about 1:2 will account for lifetimes ranging from a fraction of a second to tens of thousands of years.

But let me mention figures, for future reference. The ratios W/kT mentioned by way of example on p. 51, viz.

$$\frac{W}{kT} = 30, 50, 60,$$

producing lifetimes of

$\frac{1}{10}$s., 16 months, 30,000 years,

respectively, correspond at room temperature with threshold values of

$$0{\cdot}9,\ 1{\cdot}5,\ 1{\cdot}8\ \text{electron-volts.}$$

We must explain the unit 'electron-volt', which is rather convenient for the physicist, because it can be visualized. For example, the third number (1·8) means that an electron, accelerated by a voltage of about 2 volts, would have acquired just sufficient energy to effect the transition by impact. (For comparison, the battery of an ordinary pocket flash-light has 3 volts.)

These considerations make it conceivable that an isomeric change of configuration in some part of our molecule, produced by a chance fluctuation of the vibrational energy, can actually be a sufficiently rare event to be interpreted as a spontaneous mutation. Thus we account, by the very principles of quantum mechanics, for the most amazing fact about mutations, the fact by which they first attracted de Vries's attention, namely, that they are 'jumping' variations, no intermediate forms occurring.

STABILITY OF NATURALLY SELECTED GENES

Having discovered the increase of the natural mutation rate by any kind of ionizing rays, one might think of attributing the natural rate to the radio-activity of the soil and air and to cosmic radiation. But a quantitative comparison with the X-ray results shows that the 'natural radiation' is much too weak and could account only for a small fraction of the natural rate.

Granted that we have to account for the rare natural mutations by chance fluctuations of the heat motion, we must not be very much astonished that Nature has succeeded in making such a subtle choice of threshold values as is necessary to make mutation rare. For we have, earlier in these lectures, arrived at the conclusion that frequent mutations are detrimental to evolution. Individuals which, by mutation, acquire a gene configuration of insufficient stability, will have little chance of seeing their 'ultra-radical', rapidly mutating descendancy survive long. The species will be freed of them and will thus collect stable genes by natural selection.

THE SOMETIMES LOWER STABILITY OF MUTANTS

But, of course, as regards the mutants which occur in our breeding experiments and which we select, *qua* mutants, for studying their offspring, there is no reason to expect that they should all show that very high stability. For they have not yet been 'tried out' – or, if they have, they have been 'rejected' in the wild breeds – possibly for too high mutability. At any rate, we are not at all astonished to learn that actually some of these mutants do show a much higher mutability than the normal 'wild' genes.

TEMPERATURE INFLUENCES UNSTABLE GENES LESS THAN STABLE ONES

This enables us to test our mutability formula, which was

$$t = \tau e^{W/kT}.$$

(It will be remembered that t is the time of expectation for a mutation with threshold energy W.) We ask: How does t change with the temperature? We easily find from the preceding formula in good approximation the ratio of the value of t at temperature $T + 10$ to that at temperature T

$$\frac{{}^t T + 10}{{}^t T} = e^{-10W/kT^2}.$$

The exponent being now negative, the ratio is, naturally, smaller than 1. The time of expectation is diminished by raising the temperature, the mutability is increased. Now that can be tested and has been tested with the fly *Drosophila* in the range of temperature which the insects will stand. The result was, at first sight, surprising. The *low* mutability of wild genes was distinctly increased, but the comparatively *high* mutability occurring with some of the already mutated genes was not, or at any rate was much less, increased. That is just what we expect on comparing our two formulae. A large value of W/kT, which according to the first formula is required to make t large (stable gene), will, according to the second one, make for a small value of the ratio computed there, that is to say for a considerable increase of mutability with temperature. (The actual values of the ratio seem to lie between about $\frac{1}{2}$ and $\frac{1}{5}$. The reciprocal, 2·5, is what in an ordinary chemical reaction we call the van't Hoff factor.)

HOW X-RAYS PRODUCE MUTATION

Turning now to the X-ray-induced mutation rate, we have already inferred from the breeding experiments, first (from the proportionality of mutation rate, and dosage), that some single event produces the mutation; secondly (from quantitative results and from the fact that the mutation rate is determined by the integrated ionization density and independent of the wave-length), that this single event must be an ionization, or similar process, which has to take place inside a certain volume of only about 10 atomic-distances-cubed, in order to produce a specified mutation. According to our picture, the energy for overcoming the threshold must obviously be furnished by that explosion-like process, ionization or excitation. I call it explosion-like, because the energy spent in one ionization (spent, incidentally, not by the X-ray itself, but by a secondary electron it produces) is well known and has the comparatively enormous amount of 30 electron-volts. It is bound to be turned into enormously increased heat motion around the point where it is discharged and to spread from there in the form of a 'heat wave', a wave of intense oscillations of the atoms. That this heat wave should still be able to furnish the required threshold energy of 1 or 2 electron-volts at an average 'range of action' of about ten atomic distances, is not inconceivable, though it may well be that an unprejudiced physicist might have anticipated a slightly lower range of action. That in many cases the effect of the explosion will not be an orderly isomeric transition but a lesion of the chromosome, a lesion that becomes lethal when, by ingenious crossings, the uninjured partner (the corresponding chromosome of the second set) is removed and replaced by a partner whose corresponding gene is known to be itself morbid – all that is absolutely to be expected and it is exactly what is observed.

THEIR EFFICIENCY DOES NOT DEPEND ON SPONTANEOUS MUTABILITY

Quite a few other features are, if not predictable from the picture, easily understood from it. For example, an unstable mutant does not on the average show a much higher X-ray mutation rate than a stable one. Now, with an explosion furnishing an energy of 30 electron-volts you would certainly not expect that it makes a lot of difference whether the required threshold energy is a little larger or a little smaller, say 1 or 1·3 volts.

REVERSIBLE MUTATIONS

In some cases a transition was studied in both directions, say from a certain 'wild' gene to a specified mutant and back from that mutant to the wild gene. In such cases the natural mutation rate is sometimes nearly the same, sometimes very different. At first sight one is puzzled, because the threshold to be overcome seems to be the same in both cases. But, of course, it need not be, because it has to be measured from the energy level of the starting configuration, and that may be different for the wild and the mutated gene. (See Fig. 12 on p. 54, where '1' might refer to the wild allele, '2' to the mutant, whose lower stability would be indicated by the shorter arrow.)

On the whole, I think, Delbrück's 'model' stands the tests fairly well and we are justified in using it in further considerations.

CHAPTER 6

Order, Disorder and Entropy

Nec corpus mentem ad cogitandum, nec mens corpus ad
motum, neque ad quietem, nec ad aliquid (si quid est)
aliud determinare potest.[1] SPINOZA, *Ethics*, Pt III, Prop.2

A REMARKABLE GENERAL CONCLUSION
FROM THE MODEL

Let me refer to the phrase on p. 62, in which I tried to explain
that the molecular picture of the gene made it at least
conceivable that the miniature code should be in one-to-one
correspondence with a highly complicated and specified plan
of development and should somehow contain the means of
putting it into operation. Very well then, but how does it do
this? How are we going to turn 'conceivability' into true
understanding?

Delbrück's molecular model, in its complete generality,
seems to contain no hint as to how the hereditary substance
works. Indeed, I do not expect that any detailed information
on this question is likely to come from physics in the near
future. The advance is proceeding and will, I am sure,
continue to do so, from biochemistry under the guidance of
physiology and genetics.

No detailed information about the functioning of the genetical mechanism can emerge from a description of its structure
so general as has been given above. That is obvious. But,
strangely enough, there is just one general conclusion to be

[1]Neither can the body determine the mind to think, nor the mind determine the body
to motion or rest or anything else (if such there be).

obtained from it, and that, I confess, was my only motive for writing this book.

From Delbrück's general picture of the hereditary substance it emerges that living matter, while not eluding the 'laws of physics' as established up to date, is likely to involve 'other laws of physics' hitherto unknown, which, however, once they have been revealed, will form just as integral a part of this science as the former.

ORDER BASED ON ORDER

This is a rather subtle line of thought, open to misconception in more than one respect. All the remaining pages are concerned with making it clear. A preliminary insight, rough but not altogether erroneous, may be found in the following considerations:

It has been explained in chapter 1 that the laws of physics, as we know them, are statistical laws.[1] They have a lot to do with the natural tendency of things to go over into disorder.

But, to reconcile the high durability of the hereditary substance with its minute size, we had to evade the tendency to disorder by 'inventing the molecule', in fact, an unusually large molecule which has to be a masterpiece of highly differentiated order, safeguarded by the conjuring rod of quantum theory. The laws of chance are not invalidated by this 'invention', but their outcome is modified. The physicist is familiar with the fact that the classical laws of physics are modified by quantum theory, especially at low temperature. There are many instances of this. Life seems to be one of them, a particularly striking one. Life seems to be orderly and lawful behaviour of matter, not based exclusively on its tendency to go over from order to disorder, but based partly on existing order that is kept up.

To the physicist – but only to him – I could hope to make my view clearer by saying: The living organism seems to be a macroscopic system which in part of its behaviour approaches

---

[1] To state this in complete generality about 'the laws of physics' is perhaps challengeable. The point will be discussed in chapter 7.

to that purely mechanical (as contrasted with thermodynami-cal) conduct to which all systems tend, as the temperature approaches the absolute zero and the molecular disorder is removed.

The non-physicist finds it hard to believe that really the ordinary laws of physics, which he regards as the prototype of inviolable precision, should be based on the statistical ten-dency of matter to go over into disorder. I have given examples in chapter 1. The general principle involved is the famous Second Law of Thermodynamics (entropy principle) and its equally famous statistical foundation. On pp. 69–74 I will try to sketch the bearing of the entropy principle on the large-scale behaviour of a living organism – forgetting at the moment all that is known about chromosomes, inheritance, and so on.

LIVING MATTER EVADES THE DECAY TO EQUILIBRIUM

What is the characteristic feature of life? When is a piece of matter said to be alive? When it goes on 'doing something', moving, exchanging material with its environment, and so forth, and that for a much longer period than we would expect an inanimate piece of matter to 'keep going' under similar circumstances. When a system that is not alive is isolated or placed in a uniform environment, all motion usually comes to a standstill very soon as a result of various kinds of friction; differences of electric or chemical potential are equalized, substances which tend to form a chemical compound do so, temperature becomes uniform by heat conduction. After that the whole system fades away into a dead, inert lump of matter. A permanent state is reached, in which no observable events occur. The physicist calls this the state of thermodynamical equilibrium, or of 'maximum entropy'.

Practically, a state of this kind is usually reached very rapidly. Theoretically, it is very often not yet an absolute equilibrium, not yet the true maximum of entropy. But then the final approach to equilibrium is very slow. It could take anything between hours, years, centuries, . . . To give an example – one in which the approach is still fairly rapid: if a glass filled with pure water and a second one filled with sugared water are placed together in a hermetically closed case at constant temperature, it appears at first that nothing happens, and the impression of complete equilibrium is created. But after a day or so it is noticed that the pure water, owing to its higher vapour pressure, slowly evaporates and condenses on the solution. The latter overflows. Only after the pure water has totally evaporated has the sugar reached its aim of being equally distributed among all the liquid water available.

These ultimate slow approaches to equilibrium could never be mistaken for life, and we may disregard them here. I have referred to them in order to clear myself of a charge of inaccuracy.

IT FEEDS ON 'NEGATIVE ENTROPY'

It is by avoiding the rapid decay into the inert state of 'equilibrium' that an organism appears so enigmatic; so much so, that from the earliest times of human thought some special non-physical or supernatural force (*vis viva*, entelechy) was claimed to be operative in the organism, and in some quarters is still claimed.

How does the living organism avoid decay? The obvious answer is: By eating, drinking, breathing and (in the case of plants) assimilating. The technical term is *metabolism*. The Greek word (μεταβάλλειν) means change or exchange. Exchange of what? Originally the underlying idea is, no doubt, exchange of material. (E.g. the German for metabolism is *Stoffwechsel*.) That the exchange of material should be the essential thing is absurd. Any atom of nitrogen, oxygen, sulphur, etc., is as good as any other of its kind; what could be gained by exchanging them? For a while in the past our curiosity was silenced by being told that we feed upon energy. In some very advanced country (I don't remember whether it was Germany or the U.S.A. or both) you could find menu cards in restaurants indicating, in addition to the price, the energy content of every dish. Needless to say, taken literally, this is just as absurd. For an adult organism the energy content is as stationary as the material content. Since, surely, any calorie is worth as much as any other calorie, one cannot see how a mere exchange could help.

What then is that precious something contained in our food which keeps us from death? That is easily answered. Every process, event, happening – call it what you will; in a word, everything that is going on in Nature means an increase of the entropy of the part of the world where it is going on. Thus a living organism continually increases its entropy – or, as you may say, produces positive entropy – and thus tends to approach the dangerous state of maximum entropy, which is death. It can only keep aloof from it, i.e. alive, by continually drawing from its environment negative entropy – which is something very positive as we shall immediately see. What an organism feeds upon is negative entropy. Or, to put it less paradoxically, the essential thing in metabolism is that the organism succeeds in freeing itself from all the entropy it cannot help producing while alive.

WHAT IS ENTROPY?

What is entropy? Let me first emphasize that it is not a hazy concept or idea, but a measurable physical quantity just like the length of a rod, the temperature at any point of a body, the heat of fusion of a given crystal or the specific heat of any given substance. At the absolute zero point of temperature (roughly $-273°C$) the entropy of any substance is zero. When you bring the substance into any other state by slow, reversible little steps (even if thereby the substance changes its physical or chemical nature or splits up into two or more parts of different physical or chemical nature) the entropy increases by an amount which is computed by dividing every little portion of heat you had to supply in that procedure by the absolute temperature at which it was supplied – and by summing up all these small contributions. To give an example, when you melt a solid, its entropy increases by the amount of the heat of fusion divided by the temperature at the melting-point. You see from this, that the unit in which entropy is measured is cal./°C (just as the calorie is the unit of heat or the centimetre the unit of length).

THE STATISTICAL MEANING OF ENTROPY

I have mentioned this technical definition simply in order to remove entropy from the atmosphere of hazy mystery that frequently veils it. Much more important for us here is the bearing on the statistical concept of order and disorder, a connection that was revealed by the investigations of Boltzmann and Gibbs in statistical physics. This too is an exact quantitative connection, and is expressed by

$$\text{entropy} = k \log D,$$

where k is the so-called Boltzmann constant ($= 3 \cdot 2983 . 10^{-24}$ cal./°C), and D a quantitative measure of the atomistic disorder of the body in question. To give an exact explanation of this quantity D in brief non-technical terms is well-nigh impossible. The disorder it indicates is partly that of heat motion, partly that which consists in different kinds of atoms or molecules being mixed at random, instead of being neatly separated, e.g. the sugar and water molecules in the example quoted above. Boltzmann's equation is well illustrated by that example. The gradual 'spreading out' of the sugar over all the water available increases the disorder D, and hence (since the logarithm of D increases with D) the entropy. It is also pretty clear that any supply of heat increases the turmoil of heat motion, that is to say, increases D and thus increases the entropy; it is particularly clear that this should be so when you melt a crystal, since you thereby destroy the neat and permanent arrangement of the atoms or molecules and turn the crystal lattice into a continually changing random distribution.

An isolated system or a system in a uniform environment (which for the present consideration we do best to include as a part of the system we contemplate) increases its entropy and more or less rapidly approaches the inert state of maximum entropy. We now recognize this fundamental law of physics to be just the natural tendency of things to approach the chaotic state (the same tendency that the books of a library or the piles of papers and manuscripts on a writing desk display) unless we obviate it. (The analogue of irregular heat motion, in this case, is our handling those objects now and again without troubling to put them back in their proper places.)

ORGANIZATION MAINTAINED BY EXTRACTING 'ORDER' FROM THE ENVIRONMENT

How would we express in terms of the statistical theory the marvellous faculty of a living organism, by which it delays the decay into thermodynamical equilibrium (death)? We said before: 'It feeds upon negative entropy', attracting, as it were, a stream of negative entropy upon itself, to compensate the entropy increase it produces by living and thus to maintain itself on a stationary and fairly low entropy level.

If D is a measure of disorder, its reciprocal, $1/D$, can be regarded as a direct measure of order. Since the logarithm of $1/D$ is just minus the logarithm of D, we can write Boltzmann's equation thus:

$$- \text{(entropy)} = k \log (1/D).$$

Hence the awkward expression 'negative entropy' can be replaced by a better one: entropy, taken with the negative sign, is itself a measure of order. Thus the device by which an organism maintains itself stationary at a fairly high level of orderliness (= fairly low level of entropy) really consists in continually sucking orderliness from its environment. This conclusion is less paradoxical than it appears at first sight. Rather could it be blamed for triviality. Indeed, in the case of higher animals we know the kind of orderliness they feed upon well enough, viz. the extremely well-ordered state of matter in more or less complicated organic compounds, which serve them as foodstuffs. After utilizing it they return it in a very much degraded form – not entirely degraded, however, for plants can still make use of it. (These, of course, have their most powerful supply of 'negative entropy' in the sunlight.)

NOTE TO CHAPTER 6

The remarks on *negative entropy* have met with doubt and opposition from physicist colleagues. Let me say first, that if I had been catering for them alone I should have let the discussion turn on *free energy* instead. It is the more familiar notion in this context. But this highly technical term seemed linguistically too near to *energy* for making the average reader alive to the contrast between the two things. He is likely to take *free* as more or less an *epitheton ornans* without much relevance, while actually the concept is a rather intricate one, whose relation to Boltzmann's order–disorder principle is less easy to trace than for entropy and 'entropy taken with a negative sign', which by the way is not my invention. It happens to be precisely the thing on which Boltzmann's original argument turned.

But F. Simon has very pertinently pointed out to me that my simple thermodynamical considerations cannot account for our having to feed on matter 'in the extremely well ordered state of more or less complicated organic compounds' rather than on charcoal or diamond pulp. He is right. But to the lay reader I must explain that a piece of un-burnt coal or diamond, together with the amount of oxygen needed for its combustion, is also in an extremely well ordered state, as the physicist understands it. Witness to this: if you allow the reaction, the burning of the coal, to take place, a great amount of heat is produced. By giving it off to the surroundings, the system disposes of the very considerable entropy increase entailed by the reaction, and reaches a state in which it has, in point of fact, roughly the same entropy as before.

Yet we could not feed on the carbon dioxide that results from the reaction. And so Simon is quite right in pointing out to me, as he did, that actually the energy content of our food *does* matter; so my mocking at the menu cards that indicate it was out of place. Energy is needed to replace not only the mechanical energy of our bodily exertions, but also the heat we continually give off to the environment. And that we give off heat is not accidental, but essential. For this is precisely the manner in which we dispose of the surplus entropy we continually produce in our physical life process.

This seems to suggest that the higher temperature of the warm-blooded animal includes the advantage of enabling it to get rid of its entropy at a quicker rate, so that it can afford a more intense life process. I am not sure how much truth there is in this argument (for which I am responsible, not Simon). One may hold against it, that on the other hand many warm-blooders are *protected* against the rapid loss of heat by coats of fur or feathers. So the parallelism between body temperature and 'intensity of life', which I believe to exist, may have to be accounted for more directly by van't Hoff's law, mentioned on p. 65: the higher temperature itself speeds up the chemical reactions involved in living. (That it actually does, has been confirmed experimentally in species which take the temperature of the surroundings.)

CHAPTER 7

Is Life Based on the Laws of Physics?

Si un hombre nunca se contradice, será porque nunca dice nada.[1]

MIGUEL DE UNAMUNO (quoted from conversation)

NEW LAWS TO BE EXPECTED IN THE ORGANISM

What I wish to make clear in this last chapter is, in short, that from all we have learnt about the structure of living matter, we must be prepared to find it working in a manner that cannot be reduced to the ordinary laws of physics. And that not on the ground that there is any 'new force' or what not, directing the behaviour of the single atoms within a living organism, but because the construction is different from anything we have yet tested in the physical laboratory. To put it crudely, an engineer, familiar with heat engines only, will, after inspecting the construction of an electric motor, be prepared to find it working along principles which he does not yet understand. He finds the copper familiar to him in kettles used here in the form of long, long wires wound in coils; the iron familiar to him in levers and bars and steam cylinders is here filling the interior of those coils of copper wire. He will be convinced that it is the same copper and the same iron, subject to the same laws of Nature, and he is right in that. The difference in construction is enough to prepare him for an entirely different way of functioning. He will not suspect that an electric motor is driven by a ghost because it is set spinning by the turn of a switch, without boiler and steam.

[1] If a man never contradicts himself, the reason must be that he virtually never says anything at all.

REVIEWING THE BIOLOGICAL SITUATION

The unfolding of events in the life cycle of an organism exhibits an admirable regularity and orderliness, unrivalled by anything we meet with in inanimate matter. We find it controlled by a supremely well-ordered group of atoms, which represent only a very small fraction of the sum total in every cell. Moreover, from the view we have formed of the mechanism of mutation we conclude that the dislocation of just a few atoms within the group of 'governing atoms' of the germ cell suffices to bring about a well-defined change in the large-scale hereditary characteristics of the organism.

These facts are easily the most interesting that science has revealed in our day. We may be inclined to find them, after all, not wholly unacceptable. An organism's astonishing gift of concentrating a 'stream of order' on itself and thus escaping the decay into atomic chaos – of 'drinking orderliness' from a suitable environment – seems to be connected with the presence of the 'aperiodic solids', the chromosome molecules, which doubtless represent the highest degree of well-ordered atomic association we know of – much higher than the ordinary periodic crystal – in virtue of the individual role every atom and every radical is playing here.

To put it briefly, we witness the event that existing order displays the power of maintaining itself and of producing orderly events. That sounds plausible enough, though in finding it plausible we, no doubt, draw on experience concerning social organization and other events which involve the activity of organisms. And so it might seem that something like a vicious circle is implied.

SUMMARIZING THE PHYSICAL SITUATION

However that may be, the point to emphasize again and again is that to the physicist the state of affairs is not only not plausible but most exciting, because it is unprecedented. Contrary to the common belief, the regular course of events, governed by the laws of physics, is never the consequence of one well-ordered configuration of atoms – not unless that configuration of atoms repeats itself a great number of times, either as in the periodic crystal or as in a liquid or in a gas composed of a great number of identical molecules.

Even when the chemist handles a very complicated molecule *in vitro* he is always faced with an enormous number of like molecules. To them his laws apply. He might tell you, for example, that one minute after he has started some particular reaction half of the molecules will have reacted, and after a second minute three-quarters of them will have done so. But whether any particular molecule, supposing you could follow its course, will be among those which have reacted or among those which are still untouched, he could not predict. That is a matter of pure chance.

This is not a purely theoretical conjecture. It is not that we can never observe the fate of a single small group of atoms or even of a single atom. We can, occasionally. But whenever we do, we find complete irregularity, co-operating to produce regularity only on the average. We have dealt with an example in chapter 1. The Brownian movement of a small particle suspended in a liquid is completely irregular. But if there are many similar particles, they will by their irregular movement give rise to the regular phenomenon of diffusion.

The disintegration of a single radioactive atom is observable (it emits a projectile which causes a visible scintillation on a fluorescent screen). But if you are given a single radioactive atom, its probable lifetime is much less certain than that of a healthy sparrow. Indeed, nothing more can be said about it than this: as long as it lives (and that may be for thousands of years) the chance of its blowing up within the next second, whether large or small, remains the same. This patent lack of individual determination nevertheless results in the exact exponential law of decay of a large number of radioactive atoms of the same kind.

THE STRIKING CONTRAST

In biology we are faced with an entirely different situation. A single group of atoms existing only in one copy produces orderly events, marvellously tuned in with each other and with the environment according to most subtle laws. I said, existing only in one copy, for after all we have the example of the egg and of the unicellular organism. In the following stages of a higher organism the copies are multiplied, that is true. But to what extent? Something like 10^{14} in a grown mammal, I understand. What is that! Only a millionth of the number of molecules in one cubic inch of air. Though comparatively bulky, by coalescing they would form but a tiny drop of liquid. And look at the way they are actually distributed. Every cell harbours just one of them (or two, if we bear in mind diploidy). Since we know the power this tiny central office has in the isolated cell, do they not resemble stations of local government dispersed through the body, communicating with each other with great ease, thanks to the code that is common to all of them?

Well, this is a fantastic description, perhaps less becoming a scientist than a poet. However, it needs no poetical imagination but only clear and sober scientific reflection to recognize that we are here obviously faced with events whose regular and lawful unfolding is guided by a 'mechanism' entirely different from the 'probability mechanism' of physics. For it is simply a fact of observation that the guiding principle in every cell is embodied in a single atomic association existing only in one copy (or sometimes two) – and a fact of observation that it results in producing events which are a paragon of orderliness. Whether we find it astonishing or whether we find it quite plausible that a small but highly organized group of atoms be capable of acting in this manner, the situation is unprecedented, it is unknown anywhere else except in living matter. The physicist and the chemist, investigating inanimate matter, have never witnessed phenomena which they had to interpret in this way. The case did not arise and so our theory does not cover it – our beautiful statistical theory of which we were so justly proud because it allowed us to look behind the curtain, to watch the magnificent order of exact physical law coming forth from atomic and molecular disorder; because it revealed that the most important, the most general, the all-embracing law of entropy increase could be understood without a special assumption *ad hoc*, for it is nothing but molecular disorder itself.

TWO WAYS OF PRODUCING ORDERLINESS

The orderliness encountered in the unfolding of life springs from a different source. It appears that there are two different 'mechanisms' by which orderly events can be produced: the 'statistical mechanism' which produces 'order from disorder' and the new one, producing 'order from order'. To the unprejudiced mind the second principle appears to be much simpler, much more plausible. No doubt it is. That is why physicists were so proud to have fallen in with the other one, the 'order-from-disorder' principle, which is actually followed in Nature and which alone conveys an understanding of the great line of natural events, in the first place of their irreversibility. But we cannot expect that the 'laws of physics' derived from it suffice straightaway to explain the behaviour of living matter, whose most striking features are visibly based to a large extent on the 'order-from-order' principle. You would not expect two entirely different mechanisms to bring about the same type of law – you would not expect your latch-key to open your neighbour's door as well.

We must therefore not be discouraged by the difficulty of interpreting life by the ordinary laws of physics. For that is just what is to be expected from the knowledge we have gained of the structure of living matter. We must be prepared to find a new type of physical law prevailing in it. Or are we to term it a non-physical, not to say a super-physical, law?

THE NEW PRINCIPLE IS NOT ALIEN TO PHYSICS

No. I do not think that. For the new principle that is involved is a genuinely physical one: it is, in my opinion, nothing else than the principle of quantum theory over again. To explain this, we have to go to some length, including a refinement, not to say an amendment, of the assertion previously made, namely, that all physical laws are based on statistics.

This assertion, made again and again, could not fail to arouse contradiction. For, indeed, there are phenomena whose conspicuous features are visibly based directly on the 'order-from-order' principle and appear to have nothing to do with statistics or molecular disorder.

The order of the solar system, the motion of the planets, is maintained for an almost indefinite time. The constellation of this moment is directly connected with the constellation at any particular moment in the times of the Pyramids; it can be traced back to it, or vice versa. Historical eclipses have been calculated and have been found in close agreement with historical records or have even in some cases served to correct the accepted chronology. These calculations do not imply any statistics, they are based solely on Newton's law of universal attraction.

Nor does the regular motion of a good clock or of any similar mechanism appear to have anything to do with statistics. In short, all purely mechanical events seem to follow distinctly and directly the 'order-from-order' principle. And if we say 'mechanical', the term must be taken in a wide sense. A very useful kind of clock is, as you know, based on the regular transmission of electric pulses from the power station.

I remember an interesting little paper by Max Planck on the topic 'The Dynamical and the Statistical Type of Law' ('Dynamische und Statistische Gesetzmässigkeit'). The distinction is precisely the one we have here labelled as 'order from order' and 'order from disorder'. The object of that paper was to show how the interesting statistical type of law, controlling large-scale events, is constituted from the 'dynamical' laws supposed to govern the small-scale events, the interaction of the single atoms and molecules. The latter type is illustrated by large-scale mechanical phenomena, as the motion of the planets or of a clock, etc.

Thus it would appear that the 'new' principle, the order-from-order principle, to which we have pointed with great solemnity as being the real clue to the understanding of life, is not at all new to physics. Planck's attitude even vindicates priority for it. We seem to arrive at the ridiculous conclusion that the clue to the understanding of life is that it is based on a pure mechanism, a 'clock-work' in the sense of Planck's paper. The conclusion is not ridiculous and is, in my opinion, not entirely wrong, but it has to be taken 'with a very big grain of salt'.

THE MOTION OF A CLOCK

Let us analyse the motion of a real clock accurately. It is not at all a purely mechanical phenomenon. A purely mechanical clock would need no spring, no winding. Once set in motion, it would go on for ever. A real clock without a spring stops after a few beats of the pendulum, its mechanical energy is turned into heat. This is an infinitely complicated atomistic process. The general picture the physicist forms of it compels him to admit that the inverse process is not entirely impossible: a springless clock might suddenly begin to move, at the expense of the heat energy of its own cog wheels and of the environ-ment. The physicist would have to say: The clock experiences an exceptionally intense fit of Brownian movement. We have seen in chapter 2 (p. 16) that with a very sensitive torsional balance (electrometer or galvanometer) that sort of thing happens all the time. In the case of a clock it is, of course, infinitely unlikely.

Whether the motion of a clock is to be assigned to the dynamical or to the statistical type of lawful events (to use Planck's expressions) depends on our attitude. In calling it a dynamical phenomenon we fix attention on the regular going that can be secured by a comparatively weak spring, which overcomes the small disturbances by heat motion, so that we may disregard them. But if we remember that without a spring the clock is gradually slowed down by friction, we find that this process can only be understood as a statistical phenomenon.

However insignificant the frictional and heating effects in a clock may be from the practical point of view, there can be no doubt that the second attitude, which does not neglect them, is the more fundamental one, even when we are faced with the regular motion of a clock that is driven by a spring. For it must not be believed that the driving mechanism really does away with the statistical nature of the process. The true physical picture includes the possibility that even a regularly going clock should all at once invert its motion and, working backward, rewind its own spring – at the expense of the heat of the environment. The event is just 'still a little less likely' than a 'Brownian fit' of a clock without driving mechanism.

CLOCKWORK AFTER ALL STATISTICAL

Let us now review the situation. The 'simple' case we have analysed is representative of many others – in fact of all such as appear to evade the all-embracing principle of molecular statistics. Clockworks made of real physical matter (in contrast to imagination) are not true 'clock-works'. The element of chance may be more or less reduced, the likelihood of the clock suddenly going altogether wrong may be infinitesimal, but it always remains in the background. Even in the motion of the celestial bodies irreversible frictional and thermal influences are not wanting. Thus the rotation of the earth is slowly diminished by tidal friction, and along with this reduction the moon gradually recedes from the earth, which would not happen if the earth were a completely rigid rotating sphere.

Nevertheless the fact remains that 'physical clock-works' visibly display very prominent 'order-from-order' features – the type that aroused the physicist's excitement when he encountered them in the organism. It seems likely that the two cases have after all something in common. It remains to be seen what this is and what is the striking difference which makes the case of the organism after all novel and unprecedented.

NERNST'S THEOREM

When does a physical system – any kind of association of atoms – display 'dynamical law' (in Planck's meaning) or 'clock-work features'? Quantum theory has a very short answer to this question, viz. at the absolute zero of temperature. As zero temperature is approached the molecular disorder ceases to have any bearing on physical events. This fact was, by the way, not discovered by theory, but by carefully investigating chemical reactions over a wide range of temperatures and extrapolating the results to zero temperature – which cannot actually be reached. This is Walther Nernst's famous 'Heat Theorem', which is sometimes, and not unduly, given the proud name of the 'Third Law of Thermodynamics' (the first being the energy principle, the second the entropy principle).

Quantum theory provides the rational foundation of Nernst's empirical law, and also enables us to estimate how closely a system must approach to the absolute zero in order to display an approximately 'dynamical' behaviour. What temperature is in any particular case already practically equivalent to zero?

Now you must not believe that this always has to be a very low temperature. Indeed, Nernst's discovery was induced by the fact that even at room temperature entropy plays an astonishingly insignificant role in many chemical reactions. (Let me recall that entropy is a direct measure of molecular disorder, viz. its logarithm.)

THE PENDULUM CLOCK IS VIRTUALLY AT ZERO TEMPERATURE

What about a pendulum clock? For a pendulum clock room temperature is practically equivalent to zero. That is the reason why it works 'dynamically'. It will continue to work as it does if you cool it (provided that you have removed all traces of oil!). But it does not continue to work if you heat it above room temperature, for it will eventually melt.

THE RELATION BETWEEN CLOCKWORK AND ORGANISM

That seems very trivial but it does, I think, hit the cardinal point. Clockworks are capable of functioning 'dynamically', because they are built of solids, which are kept in shape by London–Heitler forces, strong enough to elude the disorderly tendency of heat motion at ordinary temperature.

Now, I think, few words more are needed to disclose the point of resemblance between a clockwork and an organism. It is simply and solely that the latter also hinges upon a solid – the aperiodic crystal forming the hereditary substance, largely withdrawn from the disorder of heat motion. But please do not accuse me of calling the chromosome fibres just the 'cogs of the organic machine' – at least not without a reference to the profound physical theories on which the simile is based.

For, indeed, it needs still less rhetoric to recall the fundamental difference between the two and to justify the epithets novel and unprecedented in the biological case.

The most striking features are: first, the curious distribution of the cogs in a many-celled organism, for which I may refer to the somewhat poetical description on p. 79; and secondly, the fact that the single cog is not of coarse human make, but is the finest masterpiece ever achieved along the lines of the Lord's quantum mechanics.

On Determinism and Free Will

As a reward for the serious trouble I have taken to expound the purely scientific aspects of our problem *sine ira et studio*, I beg leave to add my own, necessarily subjective, view of the philosophical implications.

According to the evidence put forward in the preceding pages the space-time events in the body of a living being which correspond to the activity of its mind, to its self-conscious or any other actions, are (considering also their complex structure and the accepted statistical explanation of physico-chemistry) if not strictly deterministic at any rate statistico-deterministic. To the physicist I wish to emphasize that in my opinion, and contrary to the opinion upheld in some quarters, *quantum indeterminacy* plays no biologically relevant role in them, except perhaps by enhancing their purely accidental character in such events as meiosis, natural and X-ray-induced mutation and so on – and this is in any case obvious and well recognized.

For the sake of argument, let me regard this as a fact, as I believe every unbiased biologist would, if there were not the well-known, unpleasant feeling about 'declaring oneself to be a pure mechanism'. For it is deemed to contradict Free Will as warranted by direct introspection.

But immediate experiences in themselves, however various and disparate they be, are logically incapable of contradicting each other. So let us see whether we cannot draw the correct, non-contradictory conclusion from the following two premises:

(i) My body functions as a pure mechanism according to the Laws of Nature.

86

(ii) Yet I know, by incontrovertible direct experience, that I am directing its motions, of which I foresee the effects, that may be fateful and all-important, in which case I feel and take full responsibility for them.

The only possible inference from these two facts is, I think, that I – I in the widest meaning of the word, that is to say, every conscious mind that has ever said or felt 'I' – am the person, if any, who controls the 'motion of the atoms' according to the Laws of Nature.

Within a cultural milieu (*Kulturkreis*) where certain conceptions (which once had or still have a wider meaning amongst other peoples) have been limited and specialized, it is daring to give to this conclusion the simple wording that it requires. In Christian terminology to say: 'Hence I am God Almighty' sounds both blasphemous and lunatic. But please disregard these connotations for the moment and consider whether the above inference is not the closest a biologist can get to proving God and immortality at one stroke.

In itself, the insight is not new. The earliest records to my knowledge date back some 2,500 years or more. From the early great Upanishads the recognition ATHMAN = BRAHMAN (the personal self equals the omnipresent, all-comprehending eternal self) was in Indian thought considered, far from being blasphemous, to represent the quintessence of deepest insight into the happenings of the world. The striving of all the scholars of Vedanta was, after having learnt to pronounce with their lips, really to assimilate in their minds this grandest of all thoughts.

Again, the mystics of many centuries, independently, yet in perfect harmony with each other (somewhat like the particles in an ideal gas) have described, each of them, the unique experience of his or her life in terms that can be condensed in the phrase: DEUS FACTUS SUM (I have become God).

To Western ideology the thought has remained a stranger, in spite of Schopenhauer and others who stood for it and in spite of those true lovers who, as they look into each other's eyes, become aware that their thought and their joy are *numerically* one – not merely similar or identical; but they, as a rule, are emotionally too busy to indulge in clear thinking, in which respect they very much resemble the mystic.

Allow me a few further comments. Consciousness is never experienced in the plural, only in the singular. Even in the pathological cases of split consciousness or double personality the two persons alternate, they are never manifest simultaneously. In a dream we do perform several characters at the same time, but not indiscriminately: we *are* one of them; in him we act and speak directly, while we often eagerly await the answer or response of another person, unaware of the fact that it is we who control his movements and his speech just as much as our own.

How does the idea of plurality (so emphatically opposed by the Upanishad writers) arise at all? Consciousness finds itself intimately connected with, and dependent on, the physical state of a limited region of matter, the body. (Consider the changes of mind during the development of the body, as puberty, ageing, dotage, etc., or consider the effects of fever, intoxication, narcosis, lesion of the brain and so on.) Now, there is a great plurality of similar bodies. Hence the pluralization of consciousnesses or minds seems a very suggestive hypothesis. Probably all simple, ingenuous people, as well as the great majority of Western philosophers, have accepted it.

It leads almost immediately to the invention of souls, as many as there are bodies, and to the question whether they are mortal as the body is or whether they are immortal and capable of existing by themselves. The former alternative is distasteful, while the latter frankly forgets, ignores or disowns the facts upon which the plurality hypothesis rests. Much sillier questions have been asked: Do animals also have souls? It has even been questioned whether women, or only men, have souls.

Such consequences, even if only tentative, must make us suspicious of the plurality hypothesis, which is common to all official Western creeds. Are we not inclining to much greater nonsense, if in discarding their gross superstitions we retain their naïve idea of plurality of souls, but 'remedy' it by declaring the souls to be perishable, to be annihilated with the respective bodies?

The only possible alternative is simply to keep to the immediate experience that consciousness is a singular of which the plural is unknown; that there *is* only one thing and that what seems to be a plurality is merely a series of different aspects of this one thing, produced by a deception (the Indian MAJA); the same illusion is produced in a gallery of mirrors, and in the same way Gaurisankar and Mt Everest turned out to be the same peak seen from different valleys.

There are, of course, elaborate ghost-stories fixed in our minds to hamper our acceptance of such simple recognition. E.g. it has been said that there is a tree there outside my window but I do not really see the tree. By some cunning device of which only the initial, relatively simple steps are explored, the real tree throws an image of itself into my consciousness, and that is what I perceive. If you stand by my side and look at the same tree, the latter manages to throw an image into your soul as well. I see my tree and you see yours (remarkably like mine), and what the tree in itself is we do not know. For this extravagance Kant is responsible. In the order of ideas which regards consciousness as a *singulare tantum* it is conveniently replaced by the statement that there is obviously only *one* tree and all the image business is a ghost-story.

Yet each of us has the indisputable impression that the sum total of his own experience and memory forms a unit, quite distinct from that of any other person. He refers to it as 'I'. *What is this 'I'?*

If you analyse it closely you will, I think, find that it is just a little bit more than a collection of single data (experiences and memories), namely the canvas *upon which* they are collected. And you will, on close introspection, find that what you really mean by 'I' is that ground-stuff upon which they are collected. You may come to a distant country, lose sight of all your friends, may all but forget them; you acquire new friends, you share life with them as intensely as you ever did with your old ones. Less and less important will become the fact that, while living your new life, you still recollect the old one. 'The youth that was I', you may come to speak of him in the third person, indeed the protagonist of the novel you are reading is probably nearer to your heart, certainly more intensely alive and better known to you. Yet there has been no intermediate break, no death. And even if a skilled hypnotist succeeded in blotting out entirely all your earlier reminiscences, you would not find that he had killed *you*. In no case is there a loss of personal existence to deplore.

Nor will there ever be.

NOTE TO THE EPILOGUE

The point of view taken here levels with what Aldous Huxley has recently – and very appropriately – called *The Perennial Philosophy*. His beautiful book (London, Chatto and Windus, 1946) is singularly fit to explain not only the state of affairs, but also why it is so difficult to grasp and so liable to meet with opposition.

MIND AND MATTER

The Tarner Lectures

delivered at Trinity College, Cambridge,
in October 1956

To
my famous and
beloved friend
HANS HOFF
in deep devotion

CHAPTER I

The Physical Basis of Consciousness

THE PROBLEM

The world is a construct of our sensations, perceptions, memories. It is convenient to regard it as existing objectively on its own. But it certainly does not become manifest by its mere existence. Its becoming manifest is conditional on very special goings-on in very special parts of this very world, namely on certain events that happen in a brain. That is an inordinately peculiar kind of implication, which prompts the question: What particular properties distinguish these brain processes and enable them to produce the manifestation? Can we guess which material processes have this power, which not? Or simpler: What kind of material process is directly associated with consciousness?

A rationalist may be inclined to deal curtly with this question, roughly as follows. From our own experience, and as regards the higher animals from analogy, consciousness is linked up with certain kinds of events in organized, living matter, namely, with certain nervous functions. How far back or 'down' in the animal kingdom there is still some sort of consciousness, and what it may be like in its early stages, are gratuitous speculations, questions that cannot be answered and which ought to be left to idle dreamers. It is still more gratuitous to indulge in thoughts about whether perhaps other events as well, events in inorganic matter, let alone all material events, are in some way or other associated with consciousness. All this is pure fantasy, as irrefutable as it is unprovable, and thus of no value for knowledge.

He who accepts this brushing aside of the question ought to

be told what an uncanny gap he thereby allows to remain in his picture of the world. For the turning-up of nerve cells and brains within certain strains of organisms is a very special event whose meaning and significance is quite well understood. It is a special kind of mechanism by which the individual responds to alternative situations by accordingly alternating behaviour, a mechanism for adaptation to a changing surrounding. It is the most elaborate and the most ingenious among all such mechanisms, and wherever it turns up it rapidly gains a dominating role. However, it is not *sui generis*. Large groups of organisms, in particular the plants, achieve very similar performances in an entirely different fashion.

Are we prepared to believe that this very special turn in the development of the higher animals, a turn that might after all have failed to appear, was a necessary condition for the world to flash up to itself in the light of consciousness? Would it otherwise have remained a play before empty benches, not existing for anybody, thus quite properly speaking not existing? This would seem to me the bankruptcy of a world picture. The urge to find a way out of this impasse ought not to be damped by the fear of incurring the wise rationalists' mockery.

According to Spinoza every particular thing or being is a modification of the infinite substance, i.e. of God. It expresses itself by each of his attributes, in particular that of extension and that of thought. The first is its bodily existence in space and time, the second is – in the case of a living man or animal – his mind. But to Spinoza any inanimate bodily thing is at the same time also 'a thought of God', that is, it exists in the second attribute as well. We encounter here the bold thought of universal animation, though not for the first time, not even in Western philosophy. Two thousand years earlier the Ionian philosophers acquired from it the surname of *hylozoists*. After Spinoza the genius of Gustav Theodor Fechner did not shy at attributing a soul to a plant, to the earth as a celestial body, to the planetary system, etc. I do not fall in with these fantasies, yet I should not like to have to pass judgment as to who has come nearer to the deepest truth, Fechner or the bankrupts of rationalism.

A TENTATIVE ANSWER

You see that all the attempts at extending the domain of consciousness, asking oneself whether anything of the sort might be reasonably associated with other than nervous processes, needs must run into unproved and unprovable speculation. But we tread on firmer ground when we start in the opposite direction. Not every nervous process, nay by no means every cerebral process, is accompanied by consciousness. Many of them are not, even though physiologically and biologically they are very much like the 'conscious' ones, both in frequently consisting of afferent impulses followed by efferent ones, and in their biological significance of regulating and timing reactions partly inside the system, partly towards a changing environment. In the first instance we meet here with the reflex actions in the vertebral ganglia and in that part of the nervous system which they control. But also (and this we shall make our special study) many reflexive processes exist that do pass through the brain, yet do not fall into consciousness at all or have very nearly ceased to do so. For in the latter case the distinction is not sharp; intermediate degrees between fully conscious and completely unconscious occur. By examining various representatives of physiologically very similar processes, all playing within our own body, it ought not to be too difficult to find out by observation and reasoning the distinctive characteristics we are looking for.

To my mind the key is to be found in the following well-known facts. Any succession of events in which we take part with sensations, perceptions and possibly with actions gradually drops out of the domain of consciousness when the same string of events repeats itself in the same way very often. But it is immediately shot up into the conscious region, if at such a repetition either the occasion or the environmental conditions met with on its pursuit differ from what they were on all the previous incidences. Even so, at first anyhow, only those modifications or 'differentials' intrude into the conscious sphere that distinguish the new incidence from previous ones and thereby usually call for 'new considerations'. Of all this each of us can supply dozens of examples out of personal experience, so that I may forgo enumerating any at the moment.

The gradual fading from consciousness is of outstanding importance to the entire structure of our mental life, which is wholly based on the process of acquiring practice by repetition, a process which Richard Semon has generalized to the concept of *Mneme*, about which we shall have more to say later. A single experience that is never to repeat itself is biologically irrelevant. Biological value lies only in learning the suitable reaction to a situation that offers itself again and again, in many cases periodically, and always requires the same response if the organism is to hold its ground. Now from our own inner experience we know the following. On the first few repetitions a new element turns up in the mind, the 'already met with' or 'notal' as Richard Avenarius has called it. On frequent repetition the whole string of events becomes more and more of a routine, it becomes more and more uninteresting, the responses become ever more reliable according as they fade from consciousness. The boy recites his poem, the girl plays her piano sonata 'well-nigh in their sleep'. We follow the habitual path to our workshop, cross the road at the customary places, turn into side-streets, etc., whilst our thoughts are occupied with entirely different things. But whenever the situation exhibits a relevant differential – let us say the road is up at the place where we used to cross it, so that we have to make a detour – this differential and our response to it intrude into consciousness, from which, however, they soon fade below the threshold, if the differential becomes a constantly repeated feature. Faced with changing alternatives, bifurcations develop and may be fixed in the same way. We branch off to the University Lecture Rooms or to the Physics Laboratory at the right point without much thinking, provided that both are frequently occurring destinations.

Now in this fashion differentials, variants of response, bifurcations, etc., are piled up one upon the other in unsurveyable abundance, but only the most recent ones remain in the domain of consciousness, only those with regard to which the living substance is still in the stage of learning or practising. One might say, metaphorically, that consciousness is the tutor who supervises the education of the living substance, but leaves his pupil alone to deal with all those tasks for which he is already sufficiently trained. But I wish to underline three times in red ink that I mean this only as a metaphor. The fact is only this, that new situations and the new responses they prompt are kept in the light of consciousness; old and well practised ones are no longer so.

Hundreds and hundreds of manipulations and performances of everyday life had all to be learnt once, and that with great attentiveness and painstaking care. Take for example a small child's first attempts in walking. They are eminently in the focus of consciousness; the first successes are hailed by the peformer with shouts of joy. When the adult laces his boots, switches on the light, takes off his clothes in the evening, eats with knife and fork . . ., these performances, that all had to be toilsomely learnt, do not in the least disturb him in the thoughts in which he may just be engaged. This may occasionally result in comical miscarriages. There is the story of a famous mathematician, whose wife is said to have found him lying in his bed, the lights switched off, shortly after an invited evening party had gathered in his house. What had happened? He had gone to his bedroom to put on a fresh shirt-collar. But the mere action of taking off the old collar had released in the man, deeply entrenched in thought, the string of performances that habitually followed in its wake.

Now this whole state of affairs, so well known from the *ontogeny* of our mental life, seems to me to shed light on the *phylogeny* of unconscious nervous processes, as in the heart beat, the peristalsis of the bowels, etc. Faced with nearly constant or regularly changing situations, they are very well and reliably practised and have, therefore, long ago dropped from the sphere of consciousness. Here too we find intermediate grades, for example, breathing, that usually goes on inadvertently, but may on account of differentials in the situation, say in smoky air or in an attack of asthma, become modified and conscious. Another instance is the bursting into tears for sorrow, joy or bodily pain, an event which, though conscious, can hardly be influenced by will. Also comical miscarriages of a mnemically inherited nature occur, as the bristling of the hair by terror, the ceasing of secretion of saliva on intense excitement, responses which must have had some significance in the past, but have lost it in the case of man.

I doubt whether everybody will readily agree with the next step, which consists in extending these notions to other than nervous processes. For the moment I shall only briefly hint at it, though to me personally it is the most important one. For this generalization precisely sheds light on the problem from which we started: What material events are associated with, or accompanied by, consciousness, what not? The answer that I suggest is as follows: What in the preceding we have said and shown to be a property of nervous processes is a property of organic processes in general, namely, to be associated with consciousness inasmuch as they are new.

In the notion and terminology of Richard Semon the ontogeny not only of the brain but of the whole individual soma is the 'well memorized' repetition of a string of events that have taken place in much the same fashion a thousand times before. Its first stages, as we know from our own experience, are unconscious – first in the mother's womb; but even the ensuing weeks and months of life are for the greatest part passed in sleep. During this time the infant carries on an evolution of old standing and habit, in which it meets with conditions that from case to case vary very little. The ensuing organic development begins to be accompanied by consciousness only inasmuch as there are organs that gradually take up interaction with the environment, adapt their functions to the changes in the situation, are influenced, undergo practice, are in special ways modified by the surroundings. We higher vertebrates possess such an organ mainly in our nervous system. Therefore consciousness is associated with those of its functions that adapt themselves by what we call experience to a changing environment. The nervous system is the place where our species is still engaged in phylogenetic transformation; metaphorically speaking it is the 'vegetation top' (*Vegetationsspitze*) of our stem. I would summarize my general hypothesis thus: consciousness is associated with the *learning* of the living substance; its *knowing how (Können)* is unconscious.

ETHICS

Even without this last generalization, which to me is very important but may still seem rather dubious to others, the theory of consciousness that I have adumbrated seems to pave the way towards a scientific understanding of ethics.

At all epochs and with all peoples the background of every ethical code (*Tugendlehre*) to be taken seriously has been, and is, self-denial (*Selbstüberwindung*). The teaching of ethics always assumes the form of a demand, a challenge, of a 'thou shalt', that is in some way opposed to our primitive will. Whence comes this peculiar contrast between the 'I will' and the 'thou shalt'? Is it not absurd that I am supposed to suppress my primitive appetites, disown my true self, be different from what I really am? Indeed in our days, more perhaps than in others, we hear this demand often enough mocked at. 'I am as I am, give room to my individuality! Free development to the desires that nature has planted in me! All the shalls that oppose me in this are nonsense, priests' fraud. God is Nature, and Nature may be credited with having formed me as she wants me to be.' Such slogans are heard occasionally. It is not easy to refute their plain and brutal obviousness. Kant's imperative is avowedly irrational.

But fortunately the scientific foundation of these slogans is worm-eaten. Our insight into the 'becoming' (*das Werden*) of the organisms makes it easy to understand that our conscious life – I will not say shall be, but that it actually is necessarily a continued fight against our primitive ego. For our natural self, our primitive will with its innate desires, is obviously the mental correlate of the material bequest received from our ancestors. Now as a species we are developing, and we march in the front-line of generations; thus every day of a man's life represents a small bit of the evolution of our species, which is still in full swing. It is true that a single day of one's life, nay even any individual life as a whole, is but a minute blow of the chisel at the ever unfinished statue. But the whole enormous evolution we have gone through in the past, it too has been brought about by myriads of such minute chisel blows. The material for this transformation, the presupposition for its taking place, are of course the inheritable spontaneous mutations. However, for selection among them, the behaviour of the carrier of the mutation, his habits of life, are of outstanding importance and decisive influence. Otherwise the origin of species, the ostensibly directed trends along which selection proceeds, could not be understood even in the long spaces of time which are after all limited and whose limits we know quite well.

And thus at every step, on every day of our life, as it were, something of the shape that we possessed until then has to change, to be overcome, to be deleted and replaced by something new. The resistance of our primitive will is the psychical correlate of the resistance of the existing shape to the transforming chisel. For we ourselves are chisel and statue, conquerors and conquered at the same time – it is a true continued 'self-conquering' (*Selbstüberwindung*).

But is it not absurd to suggest that this process of evolution should directly and significantly fall into consciousness, considering its inordinate slowness not only compared with the short span of an individual life, but even with historical epochs? Does it not just run along unnoticed?

No. In the light of our previous considerations this is not so. They culminated in regarding consciousness as associated with such physiological goings-on as are still being transformed by mutual interaction with a changing environment. Moreover, we concluded that only those modifications become conscious which are still in the stage of being trained, until, in a much later time, they become a hereditarily fixed, well-trained and unconscious possession of the species. In brief: consciousness is a phenomenon in the zone of evolution. This world lights up to itself only where or only inasmuch as it develops, procreates new forms. Places of stagnancy slip from consciousness; they may only appear in their interplay with places of evolution.

If this is granted it follows that consciousness and discord with one's own self are inseparably linked up, even that they must, as it were, be proportional to each other. This sounds a paradox, but the wisest of all times and peoples have testified to confirm it. Men and women for whom this world was lit in an unusually bright light of awareness, and who by life and word have, more than others, formed and transformed that work of art which we call humanity, testify by speech and writing or even by their very lives that more than others have they been torn by the pangs of inner discord. Let this be a consolation to him who also suffers from it. Without it nothing enduring has ever been begotten.

Please do not misunderstand me. I am a scientist, not a teacher of morals. Do not take it that I wish to propose the idea of our species developing towards a higher goal as an effective motive to propagate the moral code. This it cannot be, since it is an unselfish goal, a disinterested motive, and thus, to be accepted, already presupposes virtuousness. I feel as unable as anybody else to explain the 'shall' of Kant's imperative. The ethical law in its simplest general form (be unselfish!) is plainly a fact, it is there, it is agreed upon even by the vast majority of those who do not very often keep it. I regard its puzzling existence as an indication of our being in the beginning of a biological transformation from an egoistic to an altruistic general attitude, of man being about to become an *animal social*. For a solitary animal egoism is a virtue that tends to preserve and improve the species; in any kind of community it becomes a destructive vice. An animal that embarks on forming states without greatly restricting egoism will perish. Phylogenetically much older state-formers as the bees, ants and termites have given up egoism completely. However, its next stage, national egoism or briefly nationalism, is still in full swing with them. A worker bee that goes astray to the wrong hive is murdered without hesitation.

Now in man something is, so it seems, on the way that is not infrequent. Above the first modification clear traces of a second one in similar direction are noticeable long before the first is even nearly achieved. Though we are still pretty vigorous egoists, many of us begin to see that nationalism too is a vice that ought to be given up. Here perhaps something very strange may make its appearance. The second step, the pacification of the struggle of peoples, may be facilitated by the fact that the first step is far from being achieved, so that egoistic motives still have a vigorous appeal. Each one of us is threatened by the terrific new weapons of aggression and is thus induced to long for peace among the nations. If we were bees, ants or Lacedaemonian warriors, to whom personal fear does not exist and cowardice is the most shameful thing in the world, warring would go on for ever. But luckily we are only men – and cowards.

The considerations and conclusions of this chapter are, with me, of very old standing; they date back more than thirty years. I never lost sight of them, but I was seriously afraid that they might have to be rejected on the ground that they appear to be based on the 'inheritance of acquired characters', in other words on Lamarckism. This we are not inclined to accept. Yet even when rejecting the inheritance of acquired characters, in other words accepting Darwin's Theory of Evolution, we find the behaviour of the individuals of a species having a very significant influence on the trend of evolution, and thus feigning a sort of sham-Lamarckism. This is explained, and clinched by the authority of Julian Huxley, in the following chapter, which, however, was written with a slightly different problem in view, and not just to lend support to the ideas put forward above.

CHAPTER 2

The Future of Understanding[1]

A BIOLOGICAL BLIND ALLEY?

We may, I believe, regard it as extremely improbable that our understanding of the world represents any definite or final stage, a maximum or optimum in any respect. By this I do not mean merely that the continuation of our research in the various sciences, our philosophical studies and religious endeavour are likely to enhance and improve our present outlook. What we are likely to gain in this way in the next, say, two and a half millennia – estimating from what we have gained since Protagoras, Democritus and Antisthenes – is insignificant compared with what I am here alluding to. There is no reason whatever for believing that our brain is the supreme *ne plus ultra* of an organ of thought in which the world is reflected. It is more likely than not that a species could acquire a similar contraption whose corresponding imagery compares with ours as ours with that of the dog, or his in turn with that of a snail.

If this be so, then – though it is not relevant in principle – it interests us, as it were for personal reasons, whether anything of the sort could be reached on our globe by our own offspring or the offspring of some of us. The globe is all right. It is a fine young leasehold, still to run under acceptable conditions of living for about the time it took us (say 1,000 million years) to develop from the earliest beginnings into what we are now.

---

[1]The material in this chapter was first broadcast as a series of three talks in the European Service of the B.B.C. in September 1950, and subsequently included in *What is Life? and other essays* (Anchor Book A 88, Doubleday and Co., New York).

But are we ourselves all right? If one accepts the present theory of evolution – and we have no better – it might seem that we have been very nearly cut off from future development. Is there still physical evolution to be expected in man, I mean to say relevant changes in our physique that become gradually fixed as inherited features, just as our present bodily self is fixed by inheritance – genotypical changes, to use the technical term of the biologist? This question is difficult to answer. We may be approaching the end of a blind alley, we may even have reached it. This would not be an exceptional event and it would not mean that our species would have to become extinct very soon. From the geological records we know that some species or even large groups seem to have reached the end of their evolutionary possibilities a very long time ago, yet they have not died out, but have remained unchanged, or without significant change, for many millions of years. The tortoises, for instance, and the crocodiles are in this sense very old groups, relics of a far remote past; we are also told that the whole large group of insects are more or less in the same boat – and they comprise a greater number of separate species than all the rest of the animal kingdom taken together. But they have changed very little in millions of years, while the rest of the living surface of the earth has during this time undergone change beyond recognition. What barred further evolution in the insects was probably this – that they had adopted the plan (you will not misunderstand this figurative expression) – that they had adopted the plan of wearing their skeleton outside instead of inside as we do. Such an outside armour, while affording protection in addition to mechanical stability, cannot grow as the bones of a mammal do between birth and maturity. This circumstance is bound to render gradual adaptive changes in the life-history of the individual very difficult.

In the case of man several arguments seem to militate against further evolution. The spontaneous inheritable changes – now called mutations – from which, according to Darwin's theory, the 'profitable' ones are automatically selected, are as a rule only small evolutionary steps, affording, if any, only a slight advantage. That is why in Darwin's deductions an important part is attributed to the usually enormous abundance of offspring, of which only a very small fraction can possibly survive. For only thus does a small amelioration in the chance of survival seem to have a reasonable likelihood of being realized. This whole mechanism appears to be blocked in civilized man – in some respects even reversed. We are, generally speaking, not willing to see our fellow-creatures suffer and perish, and so we have gradually introduced legal and social institutions which on the one hand protect life, condemn systematic infanticide, try to help every sick or frail human being to survive, while on the other hand they have to replace the natural elimination of the less fit by keeping the offspring within the limits of the available livelihood. This is achieved partly in a direct way, by birth control, partly by preventing a considerable proportion of females from mating. Occasionally – as this generation knows all too well – the insanity of war and all the disasters and blunders that follow in its wake contribute their share to the balance. Millions of adults and children of both sexes are killed by starvation, exposure, epidemics. While in the far remote past warfare between small tribes or clans is supposed to have had a positive selection value, it seems doubtful whether it ever had in historical times, and doubtless war at present has none. It means an indiscriminate killing, just as the advances in medicine and surgery result in an indiscriminate saving of lives. While justly and diametrically opposite in our esteem, both war and medical art seem to be of no selection value whatever.

THE APPARENT GLOOM OF DARWINISM

These considerations suggest that as a developing species we have come to a standstill and have little prospect of further biological advance. Even if this were so, it need not bother us. We might survive without any biological change for millions of years, like the crocodiles and many insects. Still from a certain philosophical point of view the idea is depressing, and

I should like to try and make out a case for the contrary. To do so I must enter on a certain aspect of the theory of evolution which I find supported in Professor Julian Huxley's well-known book on Evolution,[1] an aspect which according to him is not always sufficiently appreciated by recent evolutionists.

Popular expositions of Darwin's theory are apt to lead you to a gloomy and discouraging view on account of the apparent passivity of the organism in the process of evolution. Mutations occur spontaneously in the genom – the 'hereditary substance'. We have reason to believe that they are mainly due to what the physicist calls a thermodynamic fluctuation – in other words to pure chance. The individual has not the slightest influence on the hereditary treasure it receives from its parents, nor on the one it leaves to its offspring. Mutations that occur are acted on by 'natural selection of the fittest'. This again seems to mean pure chance, since it means that a favourable mutation increases the prospect for the individual of survival and of begetting offspring, to which it transmits the mutation in question. Apart from this, its activity during its lifetime seems to be biologically irrelevant. For nothing of it has an influence on the offspring: acquired properties are not inherited. Any skill or training attained is lost, it leaves no trace, it dies with the individual, it is not transmitted. An intelligent being in this situation would find that nature, as it were, refuses his collaboration – she does all herself, dooms the individual to inactivity, indeed to nihilism.

As you know, Darwin's theory was not the first systematic theory of evolution. It was preceded by the theory of Lamarck, which rests entirely on the assumption that any new features an individual has acquired by specific surroundings or behaviour during its lifetime before procreation can be, and usually are, passed on to its progeny, if not entirely, at least in traces. Thus if an animal by living on rocky or sandy soil produced protecting calluses on the soles of its feet, this callosity would gradually become hereditary so that later generations would receive it as a free gift without the hardship of acquiring it. In the same way the strength or skill or even substantial

[1] *Evolution: A Modern Synthesis* (George Allen and Unwin, 1942).

adaptation produced in any organ by its being continually used for certain ends would not be lost, but passed on, at least partly, to the offspring. This view not only affords a very simple understanding of the amazingly elaborate and specific adaptation to environment which is so characteristic of all living creatures. It is also beautiful, elating, encouraging and invigorating. It is infinitely more attractive than the gloomy aspect of passivity apparently offered by Darwinism. An intelligent being which considers itself a link in the long chain of evolution may, under Lamarck's theory, be confident that its striving and efforts for improving its abilities, both bodily and mental, are not lost in the biological sense but form a small but integrating part of the striving of the species towards higher and ever higher perfection.

Unhappily Lamarckism is untenable. The fundamental assumption on which it rests, namely, that acquired properties can be inherited, is wrong. To the best of our knowledge they cannot. The single steps of evolution are those spontaneous and fortuitous mutations which have nothing to do with the behaviour of the individual during its lifetime. And so we appear to be thrown back on the gloomy aspect of Darwinism that I have depicted above.

BEHAVIOUR INFLUENCES SELECTION

I now wish to show you that this is not quite so. Without changing anything in the basic assumptions of Darwinism we can see that the behaviour of the individual, the way it makes use of its innate faculties, plays a relevant part, nay, plays the most relevant part in evolution. There is a very true kernel in Lamarck's view, namely that there is an irrescindable causal connection between the functioning, the actually being put to profitable use of a character – an organ, any property or ability or bodily feature – and its being developed in the course of generations, and gradually improved for the purposes for which it is profitably used. This connection, I say, between being used and being improved was a very correct cognition of Lamarck's, and it subsists in our present

Darwinistic outlook, but it is easily overlooked on viewing Darwinism superficially. The course of events is almost the same as if Lamarckism were right, only the 'mechanism' by which things happen is more complicated than Lamarck thought. The point is not very easy to explain or to grasp, and so it may be useful to summarize the result in advance. To avoid vagueness, let us think of an organ, though the feature in question might be any property, habit, device, behaviour, or even any small addition to, or modification of, such a feature. Lamarck thought that the organ (a) is used, (b) is thus improved, and (c) the improvement is transmitted to the offspring. This is wrong. We have to think that the organ (a) undergoes chance variations, (b) the profitably used ones are accumulated or at least accentuated by selection, (c) this continues from generation to generation, the selected mutations constituting a lasting improvement. The most striking simulation of Lamarckism occurs – according to Julian Huxley – when the initial variations that inaugurate the process are not true mutations, not yet of the inheritable type. Yet, if profitable, they may be accentuated by what he calls organic selection, and, so to speak, pave the way for true mutations to be immediately seized upon when they happen to turn up in the 'desirable' direction.

Let us now go into some details. The most important point is to see that a new character, or modification of a character, acquired by variation, by mutation or by mutation plus some little selection, may easily arouse the organism in relation to its environment to an activity that tends to increase the usefulness of that character and hence the 'grip' of selection on it. By possessing the new or changed character the individual may be caused to change its environment – either by actually transforming it, or by migration – or it may be caused to change its behaviour towards its environment, all this in a fashion so as strongly to reinforce the usefulness of the new character and thus to speed up its further selective improvement in the same direction.

This assertion may strike you as daring, since it seems to require purpose on the side of the individual, and even a high degree of intelligence. But I wish to make the point that my statement, while it includes, of course, the intelligent, purposeful behaviour of the higher animals, is by no means restricted to them. Let us give a few examples:

Not all the individuals of a population have exactly the same environment. Some of the flowers of a wild species happen to grow in the shadow, some in sunny spots, some in the higher ranges of a lofty mountain-slope, some in the lower parts or in the valley. A mutation – say hairy foliage – which is beneficial at higher altitudes, will be favoured by selection in the higher ranges but will be 'lost' in the valley. The effect is the same as if the hairy mutants had migrated towards an environment that will favour further mutations that occur in the same direction.

Another example: their ability to fly enables birds to build their nests high up in the trees where their young ones are less accessible to some of their enemies. Primarily those who took to it had a selectional advantage. The second step is that this kind of abode was bound to select the proficient fliers among the young ones. Thus a certain ability to fly produces a change of environment, or behaviour towards the environment, which favours an accumulation of the same ability.

The most remarkable feature among living beings is that they are divided into species which are, many of them, so incredibly specialized on quite particular, often tricky performances, on which especially they rely for survival. A zoological garden is almost a curiosity show, and would be much more so, could it include an insight into the life-history of insects. Non-specialization is the exception. The rule is specialization in peculiar studied tricks which 'nobody would think of if nature had not made them'. It is difficult to believe that they have all resulted from Darwinian 'accumulation by chance'. Whether one wants it or not, one is taken by the impression of forces or tendencies away from 'the plain and simple' in certain directions towards the complicated. The 'plain and simple' seems to represent an unstable state of affairs. A departure from it provokes forces – so it seems – towards a further departure, and in the same direction. That would be difficult to understand if the development of a particular device, mechanism, organ, useful behaviour, were produced by a long pearlstring of chance events, independent of each other, such as one is used to thinking of in terms of Darwin's original conception. Actually, I believe, only the first small start 'in a certain direction' has this structure. It produces itself circumstances which 'hammer the plastic material' – by selection – more and more systematically in the direction of the advantage gained at the outset. In metaphorical speech one might say: the species has found out in which direction its chance in life lies and pursues this path.

FEIGNED LAMARCKISM

We must try to understand in a general way, and to formulate in a non-animistic fashion, how a chance-mutation, which gives the individual a certain advantage and favours its survival in a given environment, should tend to do more than that, namely to increase the opportunities for its being profitably made use of, so as to concentrate on itself, as it were, the selective influence of the environment.

To reveal this mechanism let the environment be schematically described as an ensemble of favourable and unfavourable circumstances. Among the first are food, drink, shelter, sunlight and many others, among the latter are the dangers from other living beings (enemies), poisons and the roughness of the elements. For brevity we shall refer to the first kind as 'needs' and to the second as 'foes'. Not every need can be obtained, not every foe avoided. But a living species must have acquired a behaviour that strikes a compromise in avoiding the deadliest foes and satisfying the most urgent needs from the sources of easiest access, so that it does survive. A favourable mutation makes certain sources more easily accessible, or reduces the danger from certain foes, or both. It thereby increases the chance of survival of the individuals endowed with it, but in addition it shifts the most favourable compromise, because it changes the relative weights of those needs or foes on which it bears. Individuals which – by chance or intelligence – change their behaviour accordingly will be more favoured, and thus selected. This change of behaviour is not transmitted to the next generation by the genom, not by direct inheritance, but this does not mean that it is not transmitted. The simplest, most primitive example is afforded by our species of flowers (with a habitat along an extended mountain slope) that develops a hairy mutant. The hairy mutants, favoured mainly in the top ranges, disperse their seeds in such areas so that the next generation of 'hairies', taken as a whole, has 'climbed up the slope', as it were, 'to make better use of their favourable mutation'.

In all this one must bear in mind that as a rule the whole situation is extremely dynamic, the struggle is a very stiff one. In a fairly prolific population that, at the time, survives without appreciably increasing, the foes usually overpower the needs – individual survival is an exception. Moreover, foes and needs are frequently coupled, so that a pressing need can only be met by braving a certain foe. (For instance, the antelope has to come to the river for drink, but the lion knows the place just as well as he.) The total pattern of foes and needs is intricately interwoven. Thus a slight reduction of a certain danger by a given mutation may make a considerable difference for those mutants who brave that danger and thereby avoid others. This may result in a noticeable selection not only of the genetic feature in question but also with regard to the (intended or haphazard) skill in using it. That kind of behaviour is transmitted to the offspring by example, by learning, in a generalized sense of the word. The shift of behaviour, in turn, enhances the selective value of any further mutation in the same direction.

The effect of such a display may have great similarity with the mechanism as pictured by Lamarck. Though neither an acquired behaviour nor any physical changes that it entails are directly transmitted to the offspring, yet behaviour has an important say in the process. But the causal connection is not what Lamarck thought it to be, rather just the other way round. It is not that the behaviour changes the physique of the parents and, by physical inheritance, that of the offspring. It is the physical change in the parents that modifies – directly or indirectly, by selection – their behaviour; and this change of behaviour is, by example or teaching or even more primitively, transmitted to the progeny, along with the physical change carried by the genom. Nay, even if the physical change is not yet an inheritable one, the transmission of the induced behaviour 'by teaching' can be a highly efficient evolutionary factor, because it throws the door open to receive future inheritable mutations with a prepared readiness to make the best use of them and thus to subject them to intense selection.

GENETIC FIXATION OF HABITS AND SKILLS

One might object that what we have here described may happen occasionally, but cannot continue indefinitely to form the essential mechanism of adaptive evolution. For the change of behaviour itself is not transmitted by physical inheritance, by the hereditary substance, the chromosomes. At first, therefore, it is certainly not fixed genetically and it is difficult to see how it should ever come to be incorporated in the hereditary treasure. This is an important problem in itself. For we do know that habits are inherited as, for instance, habits of nestbuilding in the birds, the various habits of cleanliness we observe in our dogs and cats, to mention a few obvious examples. If this could not be understood along orthodox Darwinian lines, Darwinism would have to be abandoned. The question becomes of singular significance in its application to man, since we wish to infer that the striving and labouring of a man during his lifetime constitute an integrating contribution to the development of the species, in the quite proper biological sense. I believe the situation to be, briefly, as follows.

According to our assumptions the behaviour changes parallel those of the physique, first as a consequence of a chance change in the latter, but very soon directing the further selectional mechanism into definite channels, because, according as behaviour has availed itself of the first rudimentary benefits, only further mutations in the same direction have any selective value. But as (let me say) the new organ develops, behaviour becomes more and more bound up with its mere possession. Behaviour and physique merge into one. You simply cannot possess clever hands without using them for obtaining your aims, they would be in your way (as they often are to an amateur on the stage, because he has only fictitious aims). You cannot have efficient wings without attempting to fly. You cannot have a modulated organ of speech without trying to imitate the noises you hear around you. To distinguish between the possession of an organ and the urge to use it and to increase its skill by practice, to regard them as two different characteristics of the organism in question, would be an artificial distinction, made possible by an abstract language but having no counterpart in nature. We must, of course, not think that 'behaviour' after all gradually intrudes into the chromosome structure (or what not) and acquires 'loci' there. It is the new organs themselves (and they do become genetically fixed) that carry along with them the habit and the way of using them. Selection would be powerless in 'producing' a new organ if selection were not aided all along by the organism's making appropriate use of it. And this is very essential. For thus, the two things go quite parallel and are ultimately, or indeed at every stage, fixed genetically as one thing: *a used organ* – as if Lamarck were right.

It is illuminating to compare this natural process with the making of an instrument by man. At first sight there appears to be a marked contrast. If we manufacture a delicate mechanism, we should in most cases spoil it if we were impatient and tried to use it again and again long before it is finished. Nature, one is inclined to say, proceeds differently. She cannot produce a new organism and its organs otherwise than whilst they are continually used, probed, examined with regard to their efficiency. But actually this parallel is wrong. The making of a single instrument by man corresponds to ontogenesis, that is, to the growing up of a single individual from the seed to maturity. Here too interference is not welcome. The young ones must be protected, they must not be put to work before they have acquired the full strength and skill of their species. The true parallel of the evolutionary development of organisms could be illustrated, for example, by a historical exhibition of bicycles, showing how this machine gradually changed from year to year, from decade to decade, or, in the same way, of railway-engines, motor-cars, aeroplanes, typewriters, etc. Here, just as in the natural process, it is obviously essential that the machine in question should be continually used and thus improved; not literally improved by use, but by the experience gained and the alterations suggested. The bicycle, by the way, illustrates the case, mentioned before, of an old organism, which has reached the attainable perfection and has therefore pretty well ceased to undergo further changes. Still it is not about to become extinct!

DANGERS TO INTELLECTUAL EVOLUTION

Let us now return to the beginning of this chapter. We started from the question: is further biological development in man likely? Our discussion has, I believe, brought to the fore two relevant points.

The first is the biological importance of behaviour. By conforming to innate faculties as well as to the environment and by adapting itself to changes in either of these factors, behaviour, though not itself inherited, may yet speed up the process of evolution by orders of magnitude. While in plants and in the lower ranges of the animal kingdom adequate behaviour is brought about by the slow process of selection, in other words by trial and error, man's high intelligence enables him to enact it by choice. This incalculable advantage may easily outweigh his handicap of slow and comparatively scarce propagation, which is further reduced by the biologically dangerous regard not to let our offspring exceed the volume for which livelihood can be secured.

The second point, concerning the question whether biological development is still to be expected in man, is intimately connected with the first. In a way we get the full answer, namely, this will depend on us and our doing. We must not wait for things to come, believing that they are decided by irrescindable destiny. If we want it, we must do something about it. If not, not. Just as the political and social development and the sequence of historical events in general are not thrust upon us by the spinning of the Fates, but largely depend on our own doing, so our biological future, being nothing else but history on a large scale, must not be taken to be an unalterable destiny that is decided in advance by any Law of Nature. To us at any rate, who are the acting subjects in the play, it is not, even though to a superior being, watching us as we watch the birds and the ants, it might appear to be. The reason why man tends to regard history, in the narrower and in the wider sense, as a predestined happening, controlled by rules and laws that he cannot change, is very obvious. It is because every single individual feels that he by himself has very little say in the matter, unless he can put his opinions over to many others and persuade them to regulate their behaviour accordingly.

As regards the concrete behaviour necessary to secure our biological future, I will only mention one general point that I consider of primary importance. We are, I believe, at the moment in grave danger of missing the 'path to perfection'. From all that has been said, selection is an indispensable requisite for biological development. If it is entirely ruled out, development stops, nay, it may be reversed. To put it in the words of Julian Huxley: ' . . . the preponderance of degenerative (loss) mutation will result in degeneration of an organ when it becomes useless and selection is accordingly no longer acting on it to keep it up to the mark.'

Now I believe that the increasing mechanization and 'stupidization' of most manufacturing processes involve the serious danger of a general degeneration of our organ of intelligence. The more the chances in life of the clever and of the unresponsive worker are equalled out by the repression of handicraft and the spreading of tedious and boring work on the assembly line, the more will a good brain, clever hands and a sharp eye become superfluous. Indeed the unintelligent man, who naturally finds it easier to submit to the boring toil, will be favoured; he is likely to find it easier to thrive, to settle down and to beget offspring. The result may easily amount even to a negative selection as regards talents and gifts.

The hardship of modern industrial life has led to certain institutions calculated to mitigate it, such as protection of the workers against exploitation and unemployment, and many other welfare and security measures. They are duly regarded as beneficial and they have become indispensable. Still we cannot shut our eyes to the fact that, by alleviating the responsibility of the individual to look after himself and by levelling the chances of every man, they also tend to rule out the competition of talents and thus to put an efficient brake on biological evolution. I realize that this particular point is highly controversial. One may make a strong case that the care for our present welfare must override the worry about our evolutionary future. But fortunately, so I believe, they go together according to my main argument. Next to want, boredom has become the worst scourge in our lives. Instead of letting the ingenious machinery we have invented produce an increasing amount of superfluous luxury, we must plan to develop it so that it takes off human beings all the unintelligent, mechanical, 'machine-like' handling. The machine must take over the toil for which man is too good, not man the work for which the machine is too expensive, as comes to pass quite often. This will not tend to make production cheaper, but those who are engaged in it happier. There is small hope of putting this through as long as the competition between big firms and concerns all over the world prevails. But this kind of competition is as uninteresting as it is biologically worthless. Our aim should be to reinstate in its place the interesting and intelligent competition of single human beings.

CHAPTER 3

The Principle of Objectivation

Nine years ago I put forward two general principles that form the basis of the scientific method, the principle of the understandability of nature, and the principle of objectivation. Since then I have touched on this matter now and again, last time in my little book *Nature and the Greeks*.[1] I wish to deal here in detail with the second one, the objectivation. Before I say what I mean by that, let me remove a possible misunderstanding which might arise, as I came to realize from several reviews of that book, though I thought I had prevented it from the outset. It is simply this: some people seemed to think that my intention was to lay down the fundamental principles which *ought* to be at the basis of scientific method or at least which justly and rightly are at the basis of science and ought to be kept at all cost. Far from this, I only maintained and maintain that they *are* – and, by the way, as an inheritance from the ancient Greeks, from whom all our Western science and scientific thought has originated.

The misunderstanding is not very astonishing. If you hear a scientist pronounce basic principles of science, stressing two of them as particularly fundamental and of old standing, it is natural to think that he is at least strongly in favour of them and wishes to impose them. But on the other hand, you see, science never imposes anything, science *states*. Science aims at nothing but making true and adequate statements about its object. The scientist only imposes two things, namely truth and sincerity, imposes them upon himself and upon other scientists. In the present case the object is science itself, as it

[1] Cambridge University Press, 1954.

117

has developed and has become and at present is, not as it *ought* to be or *ought* to develop in future.

Now let us turn to these two principles themselves. As regards the first, 'that nature can be understood', I will say here only a few words. The most astonishing thing about it is that it had to be invented, that it was at all necessary to invent it. It stems from the Milesian School, the *physiologoi*. Since then it has remained untouched, though perhaps not always uncontaminated. The present line in physics is possibly a quite serious contamination. The uncertainty principle, the alleged lack of strict causal connection in nature, may represent a step away from it, a partial abandonment. It would be interesting to discuss this, but I set my heart here on discussing the other principle, that which I called objectivation.

By this I mean the thing that is also frequently called the 'hypothesis of the real world' around us. I maintain that it amounts to a certain simplification which we adopt in order to master the infinitely intricate problem of nature. Without being aware of it and without being rigorously systematic about it, we exclude the Subject of Cognizance from the domain of nature that we endeavour to understand. We step with our own person back into the part of an onlooker who does not belong to the world, which by this very procedure becomes an objective world. This device is veiled by the following two circumstances. First, my own body (to which my mental activity is so very directly and intimately linked) forms part of the object (the real world around me) that I construct out of my sensations, perceptions and memories. Secondly, the bodies of other people form part of this objective world. Now I have very good reasons for believing that these other bodies are also linked up with, or are, as it were, the seats of spheres of consciousness. I can have no reasonable doubt about the existence or some kind of actualness of these foreign spheres of consciousness, yet I have absolutely no direct subjective access to any of them. Hence I am inclined to take them as something objective, as forming part of the real world around me. Moreover, since there is no distinction between myself and others, but on the contrary full symmetry for all intents and purposes, I conclude that I myself also form part of this real material world around me. I so to speak put my own sentient self (which had constructed this world as a mental product) back into it – with the pandemonium of disastrous logical consequences that flow from the aforesaid chain of faulty conclusions. We shall point them out one by one; for the moment let me just mention the two most blatant antinomies due to our awareness of the fact that a moderately satisfying picture of the world has only been reached at the high price of taking ourselves out of the picture, stepping back into the role of a non-concerned observer.

The first of these antinomies is the astonishment at finding our world picture 'colourless, cold, mute'. Colour and sound, hot and cold are our immediate sensations; small wonder that they are lacking in a world model from which we have removed our own mental person.

The second is our fruitless quest for the place where mind acts on matter or vice-versa, so well known from Sir Charles Sherrington's honest search, magnificently expounded in *Man on his Nature*. The material world has only been constructed at the price of taking the self, that is, mind, out of it, removing it; mind is not part of it; obviously, therefore, it can neither act on it nor be acted on by any of its parts. (This was stated in a very brief and clear sentence by Spinoza, see p. 122.)

I wish to go into more detail about some of the points I have made. First let me quote a passage from a paper of C.G. Jung which has gratified me because it stresses the same point in quite a different context, albeit in a strongly vituperative fashion. While I continue to regard the removal of the Subject of Cognizance from the objective world picture as the high price paid for a fairly satisfactory picture, for the time being, Jung goes further and blames us for paying this ransom from an inextricably difficult situation. He says:

All science (*Wissenschaft*) however is a function of the soul, in which all knowledge is rooted. The soul is the greatest of all cosmic miracles, it is the *conditio sine qua non* of the world as an object. It is exceedingly astonishing that the Western world (apart from very rare exceptions) seems to have so little appreciation of this being so. The flood of external objects of cognizance has made the subject of all cognizance withdraw to the background, often to apparent non-existence.[1]

Of course Jung is quite right. It is also clear that he, being engaged in the science of psychology, is much more sensitive to the initial gambit in question, much more so than a physicist or a physiologist. Yet I would say that a rapid withdrawal from the position held for over 2,000 years is dangerous. We may lose everything without gaining more than some freedom in a special – though very important – domain. But here the problem is set. The relatively new science of psychology imperatively demands living-space, it makes it unavoidable to reconsider the initial gambit. This is a hard task, we shall not settle it here and now, we must be content at having pointed it out.

While here we found the psychologist Jung complaining about the exclusion of the mind, the neglect of the soul, as he terms it, in our world picture, I should now like to adduce in contrast, or perhaps rather as a supplement, some quotations of eminent representatives of the older and humbler sciences of physics and physiology, just stating the fact that 'the world of science' has become so horribly objective as to leave no room for the mind and its immediate sensations.

Some readers may remember A.S. Eddington's 'two writing desks'; one is the familiar old piece of furniture at which he is seated, resting his arms on it, the other is the scientific physical body which not only lacks all and every sensual qualities but in addition is riddled with holes; by far the greatest part of it is empty space, just nothingness, interspersed with innumerable tiny specks of something, the electrons and the nuclei whirling around, but always separated by distances at least 100,000 times their own size. After having contrasted the two in his wonderfully plastic style he summarizes thus:

[1] *Eranos Jahrbuch* (1946), p. 398.

In the world of physics we watch a shadowgraph performance of familiar life. The shadow of my elbow rests on the shadow table as the shadow ink flows over the shadow paper . . . The frank realization that physical science is concerned with a world of shadows is one of the most significant of recent advances.[1]

Please note that the very recent advance does not lie in the world of physics itself having acquired this shadowy character; it had it ever since Democritus of Abdera and even before, but we were not aware of it; we thought we were dealing with the world itself; expressions like model or picture for the conceptual constructs of science came up in the second half of the nineteenth century, and not earlier, as far as I know.

Not much later Sir Charles Sherrington published his momentous *Man on his Nature*.[2] The book is pervaded by the honest search for objective evidence of the interaction between matter and mind. I stress the epithet 'honest', because it does need a very serious and sincere endeavour to look for something which one is deeply convinced in advance cannot be found, because (in the teeth of popular belief) it does not exist. A brief summary of the result of this search is found on p. 357:

Mind, the anything perception can compass, goes therefore in our spatial world more ghostly than a ghost. Invisible, intangible, it is a thing not even of outline; it is not a 'thing'. It remains without sensual confirmation and remains without it forever.

In my own words I would express this by saying: Mind has erected the objective outside world of the natural philosopher out of its own stuff. Mind could not cope with this gigantic task otherwise than by the simplifying device of excluding itself – withdrawing from its conceptual creation. Hence the latter does not contain its creator.

I cannot convey the grandeur of Sherrington's immortal book by quoting sentences; one has to read it oneself. Still, I will mention a few of the more particularly characteristic.

Physical science . . . faces us with the impasse that mind *per se* cannot play the piano – mind *per se* cannot move a finger of a hand

[1] *The Nature of the Physical World* (Cambridge University Press, 1928), Introduction.
[2] Cambridge University Press, 1940.

(p. 222).

Then the impasse meets us. The blank of the 'how' of mind's leverage on matter. The inconsequence staggers us. Is it a misunderstanding? (p. 232).

Hold these conclusions drawn by an experimental physiologist of the twentieth century against the simple statement of the greatest philosopher of the seventeenth century: B. Spinoza (*Ethics*, Pt III, Prop. 2):

Nec corpus mentem ad cogitandum, nec mens corpus ad motum, neque ad quietem, nec ad aliquid (si quid est) aliud determinare potest.

[Neither can the body determine the mind to think, nor the mind determine the body to motion or rest or anything else (if such there be).]

The impasse *is* an impasse. Are we thus not the doers of our deeds? Yet we feel responsible for them, we are punished or praised for them, as the case may be. It is a horrible antinomy. I maintain that it cannot be solved on the level of present-day science which is still entirely engulfed in the 'exclusion principle' – without knowing it – hence the antinomy. To realize this is valuable, but it does not solve the problem. You cannot remove the 'exclusion principle' by act of parliament as it were. Scientific attitude would have to be rebuilt, science must be made anew. Care is needed.

So we are faced with the following remarkable situation. While the stuff from which our world picture is built is yielded exclusively from the sense organs as organs of the mind, so that every man's world picture is and always remains a construct of his mind and cannot be proved to have any other existence, yet the conscious mind itself remains a stranger within that construct, it has no living space in it, you can spot it nowhere in space. We do not usually realize this fact, because we have entirely taken to thinking of the personality of a human being, or for that matter also that of an animal, as located in the interior of its body. To learn that it cannot really be found there is so amazing that it meets with doubt and hesitation, we are very loath to admit it. We have got used to localizing the conscious personality inside a person's head –

I should say an inch or two behind the midpoint of the eyes. From there it gives us, as the case may be, understanding or loving or tender – or suspicious or angry looks. I wonder has it ever been noted that the eye is the only sense organ whose purely receptive character we fail to recognize in naïve thought. Reversing the actual state of affairs, we are much more inclined to think of 'rays of vision', issuing from the eye, than of the 'rays of light' that hit the eyes from outside. You quite frequently find such a 'ray of vision' represented in a drawing in a comic paper, or even in some older schematic sketch intended to illustrate an optic instrument or law, a dotted line emerging from the eye and pointing to the object, the direction being indicated by an arrowhead at the far end. – Dear reader or, or better still, dear lady reader, recall the bright, joyful eyes with which your child beams upon you when you bring him a new toy, and then let the physicist tell you that in reality nothing emerges from these eyes; in reality their only objectively detectable function is, continually to be hit by and to receive light quanta. In reality! A strange reality! Something seems to be missing in it.

It is very difficult for us to take stock of the fact that the localization of the personality, of the conscious mind, inside the body is only symbolic, just an aid for practical use. Let us, with all the knowledge we have about it, follow such a 'tender look' inside the body. We do hit there on a supremely interesting bustle or, if you like, machinery. We find millions of cells of very specialized build in an arrangement that is unsurveyably intricate but quite obviously serves a very far-reaching and highly consummate mutual communication and collaboration; a ceaseless hammering of regular electro-chemical pulses which, however, change rapidly in their configuration, being conducted from nerve cell to nerve cell, tens of thousands of contacts being opened and blocked within every split second, chemical transformations being induced and maybe other changes as yet undiscovered. All this we meet and, as the science of physiology advances, we may trust that we shall come to know more and more about it. But now let us assume that in a particular case you eventually observe several efferent bundles of pulsating currents, which issue from the brain and through long cellular protrusions (motor nerve fibres), are conducted to certain muscles of the arm, which, as a consequence, tends a hesitating, trembling hand to bid you farewell – for a long, heart-rending separation; at the same time you may find that some other pulsating bundles produce a certain glandular secretion so as to veil the poor sad eye with a crape of tears. But nowhere along this way from the eye through the central organ to the arm muscles and the tear glands – nowhere, you may be sure, however far physiology advances, will you ever meet the personality, will you ever meet the dire pain, the bewildered worry within this soul, though their reality is to you so certain as though you suffered them yourself – as in actual fact you do! The picture that physiological analysis vouchsafes to us of any other human being, be it our most intimate friend, strikingly recalls to me Edgar Allan Poe's masterly story, which I am sure many a reader remembers well; I mean *The Masque of the Red Death*. A princeling and his retinue have withdrawn to an isolated castle to escape the pestilence of the red death that rages in the land. After a week or so of retirement they arrange a great dancing feast in fancy dress and mask. One of the masks, tall, entirely veiled, clad all in red and obviously intended to represent the pestilence allegorically, makes everybody shudder, both for the wantonness of the choice and for the suspicion that it might be an intruder. At last a bold young man approaches the red mask and with a sudden jolt tears off veil and head-gear. It is found empty.

Now our skulls are not empty. But what we find there, in spite of the keen interest it arouses, is truly nothing when held against the life and the emotions of the soul.

To become aware of this may in the first moment upset one. To me it seems, on deeper thought, rather a consolation. If you have to face the body of a deceased friend whom you sorely miss, is it not soothing to realize that this body was never really the seat of his personality but only symbolically 'for practical reference'?

As an appendix to these considerations, those strongly interested in the physical sciences might wish to hear me pronounce on a line of ideas, concerning subject and object, that has been given great prominence by the prevailing school of thought in quantum physics, the protagonists being Niels Bohr, Werner Heisenberg, Max Born and others. Let me first give you a very brief description of their ideas. It runs as follows:[1]

We cannot make any factual statement about a given natural object (or physical system) without 'getting in touch' with it. This 'touch' is a real physical interaction. Even if it consists only in our 'looking at the object' the latter must be hit by light-rays and reflect them into the eye, or into some instrument of observation. This means that the object is affected by our observation. You cannot obtain any knowledge about an object while leaving it strictly isolated. The theory goes on to assert that this disturbance is neither irrelevant nor completely surveyable. Thus after any number of painstaking observations the object is left in a state of which some features (the last observed) are known, but others (those interfered with by the last observation) are not known, or not accurately known. This state of affairs is offered as an explanation why no complete, gapless description of any physical object is ever possible.

If this has to be granted – and possibly it has to be granted – then it flies in the face of the principle of understandability of nature. This in itself is no opprobrium. I told you at the outset that my two principles are not meant to be binding on science, that they only express what we had actually kept to in physical science for many, many centuries and what cannot easily be changed. Personally I do not feel sure that our present knowledge as yet vindicates the change. I consider it possible that our models can be modified in such a fashion that they do not exhibit at any moment properties that cannot in principle be observed simultaneously – models poorer in simultaneous properties but richer in adaptability to changes in the environment. However, this is an internal question of

[1] See my *Science and Humanism* (Cambridge Universty Press, 1951), p. 49.

physics, not to be decided here and now. But from the theory as explained before, from the unavoidable and unsurveyable interference of the measuring devices with the object under observation, lofty consequences of an epistemological nature have been drawn and brought to the fore, concerning the relation between subject and object. It is maintained that recent discoveries in physics have pushed forward to the mysterious boundary between the subject and the object. This boundary, so we are told, is not a sharp boundary at all. We are given to understand that we never observe an object without its being modified or tinged by our own activity in observing it. We are given to understand that under the impact of our refined methods of observation and of thinking about the results of our experiments that mysterious boundary between the subject and the object has broken down.

In order to criticize these contentions let me at first accept the time-hallowed distinction or discrimination between object and subject, as many thinkers both in olden times have accepted it and in recent times still accept it. Among the philosophers who accepted it – from Democritus of Abdera down to the 'Old Man of Königsberg' – there were few, if any who did not emphasize that all our sensations, perceptions and observations have a strong, personal, subjective tinge and do not convey the nature of the 'thing-in-itself', to use Kant's term. While some of these thinkers might have in mind only a more or less strong or slight distortion, Kant landed us with a complete resignation: never to know anything at all about his 'thing-in-itself'. Thus the idea of subjectivity in all appearance is very old and familiar. What is new in the present setting is this: that not only would the impressions we get from our environment largely depend on the nature and the contingent state of our sensorium, but inversely the very environment that we wish to take in is modified by us, notably by the devices we set up in order to observe it.

Maybe this is so – to some extent it certainly is. May be that from the newly discovered laws of quantum physics this modification cannot be reduced below certain well-ascertained limits. Still I would not like to call this a direct influence of the subject on the object. For the subject, if anything, is the thing that senses and thinks. Sensations and thoughts do not belong to the 'world of energy', they cannot produce any change in this world of energy as we know from Spinoza and Sir Charles Sherrington.

All this was said from the point of view that we accept the time-hallowed discrimination between subject and object. Though we have to accept it in everyday life 'for practical reference', we ought, so I believe, to abandon it in philosophical thought. Its rigid logical consequence has been revealed by Kant: the sublime, but empty, idea of the 'thing-in-itself' about which we forever know nothing.

It is the same elements that go to compose my mind and the world. This situation is the same for every mind and its world, in spite of the unfathomable abundance of 'cross-references' between them. The world is given to me only once, not one existing and one perceived. Subject and object are only one. The barrier between them cannot be said to have broken down as a result of recent experience in the physical sciences, for this barrier does not exist.

CHAPTER 4

The Arithmetical Paradox:
The Oneness of Mind

The reason why our sentient, percipient and thinking ego is met nowhere within our scientific world picture can easily be indicated in seven words: because it is itself that world picture. It is identical with the whole and therefore cannot be contained in it as a part of it. But, of course, here we knock against the arithmetical paradox; there appears to be a great multitude of these conscious egos, the world however is only one. This comes from the fashion in which the world-concept produces itself. The several domains of 'private' consciousnesses partly overlap. The region common to all where they all overlap is the construct of the 'real world around us'. With all that an uncomfortable feeling remains, prompting such questions as: Is my world really the same as yours? Is there *one* real world to be distinguished from its pictures introjected by way of perception into every one of us? And if so, are these pictures like unto the real world or is the latter, the world 'in itself', perhaps very different from the one we perceive?

Such questions are ingenious, but in my opinion very apt to confuse the issue. They have no adequate answers. They all are, or lead to, antinomies springing from the one source, which I called the arithmetical paradox; the *many* conscious egos from whose mental experiences the *one* world is concocted. The solution of this paradox of numbers would do away with all the questions of the aforesaid kind and reveal them, I dare say, as sham questions.

There are two ways out of the number paradox, both appearing rather lunatic from the point of view of present scientific thought (based on ancient Greek thought and thus thoroughly 'Western'). One way out is the multiplication of the world in Leibniz's fearful doctrine of monads: every monad to be a world by itself, no communication between them; the monad 'has no windows', it is 'incommunicado'. That none the less they all agree with each other is called 'pre-established harmony'. I think there are few to whom this suggestion appeals, nay who would consider it as a mitigation at all of the numerical antinomy.

There is obviously only one alternative, namely the unification of minds or consciousnesses. Their multiplicity is only apparent, in truth there is only one mind. This is the doctrine of the Upanishads. And not only of the Upanishads. The mystically experienced union with God regularly entails this attitude unless it is opposed by strong existing prejudices; and this means that it is less easily accepted in the West than in the East. Let me quote as an example outside the Upanishads an Islamic Persian mystic of the thirteenth century, Aziz Nasafi. I am taking it from a paper by Fritz Meyer[1] and translating from his German translation:

On the death of any living creature the spirit returns to the spiritual world, the body to the bodily world. In this however only the bodies are subject to change. The spiritual world is one single spirit who stands like unto a light behind the bodily world and who, when any single creature comes into being, shines through it as through a window. According to the kind and size of the window less or more light enters the world. The light itself however remains unchanged.

Ten years ago Aldous Huxley published a precious volume which he called *The Perennial Philosophy*[2] and which is an anthology from the mystics of the most various periods and the most various peoples. Open it where you will and you find many beautiful utterances of a similar kind. You are struck by the miraculous agreement between humans of different race, different religion, knowing nothing about each other's existence, separated by centuries and millennia, and by the greatest distances that there are on our globe.

Still, it must be said that to Western thought this doctrine

[1] *Eranos Jahrbuch*, 1946.
[2] Chatto and Windus, 1946.

has little appeal, it is unpalatable, it is dubbed fantastic, unscientific. Well, so it is because our science – Greek science – is based on objectivation, whereby it has cut itself off from an adequate understanding of the Subject of Cognizance, of the mind. But I do believe that this is precisely the point where our present way of thinking does need to be amended, perhaps by a bit of blood-transfusion from Eastern thought. That will not be easy, we must beware of blunders – blood-transfusion always needs great precaution to prevent clotting. We do not wish to lose the logical precision that our scientific thought has reached, and that is unparalleled anywhere at any epoch.

Still, one thing can be claimed in favour of the mystical teaching of the 'identity' of all minds with each other and with the supreme mind – as against the fearful monadology of Leibniz. The doctrine of identity can claim that it is clinched by the empirical fact that consciousness is never experienced in the plural, only in the singular. Not only has none of us ever experienced more than one consciousness, but there is also no trace of circumstantial evidence of this ever happening anywhere in the world. If I say that there cannot be more than one consciousness in the same mind, this seems a blunt tautology – we are quite unable to imagine the contrary.

Yet there are cases or situations where we would expect and nearly require this unimaginable thing to happen, if it can happen at all. This is the point that I should like to discuss now in some detail, and to clinch it by quotations from Sir Charles Sherrington, who was at the same time (rare event!) a man of highest genius and a sober scientist. For all I know he had no bias towards the philosophy of the Upanishads. My purpose in this discussion is to contribute perhaps to clearing the way for a future assimilation of the doctrine of identity with our own scientific world view, without having to pay for it by a loss of soberness and logical precision.

I said just now that we are not able even to imagine a plurality of consciousnesses in one mind. We can pronounce these words all right, but they are not the description of any thinkable experience. Even in the pathological cases of a 'split personality' the two persons alternate, they never hold the field jointly; nay this is just the characteristic feature, that they know nothing about each other.

When in the puppet-show of dream we hold in hand the strings of quite a number of actors, controlling their actions and their speech, we are not aware of this being so. Only one of them is myself, the dreamer. In him I act and speak immediately, while I may be awaiting eagerly and anxiously what another one will reply, whether he is going to fulfil my urgent request. That I could really let him do and say whatever I please does not occur to me – in fact it is not quite the case. For in a dream of this kind the 'other one' is, I dare say, mostly the impersonation of some serious obstacle that opposes me in waking life and of which I have actually no control. The strange state of affairs, described here, is quite obviously the reason why most people of old firmly believed that they were truly in communication with the persons, alive or deceased, or, maybe, gods or heroes, whom they met in their dreams. It is a superstition that dies hard. On the verge of the sixth century B.C. Heraclitus of Ephesus definitely pronounced against it, with a clarity not often met with in his sometimes very obscure fragments. But Lucretius Carus, who believed himself to be the protagonist of enlightened thought, still holds on to this superstition in the first century B.C. In our days it is probably rare, but I doubt that it is entirely extinct.

Let me turn to something quite different. I find it utterly impossible to form an idea about either how, for example, my own conscious mind (that I feel to be *one*) should have originated by integration of the consciousnesses of the cells (or some of them) that form my body, or how it should at every moment of my life be, as it were, their resultant. One would think that such a 'commonwealth of cells' as each of us is would be the occasion *par excellence* for mind to exhibit plurality if it were at all able to do so. The expression 'commonwealth' or 'state of cells' (*Zellstaat*) is nowadays no longer to be regarded as a metaphor. Listen to Sherrington:

To declare that, of the component cells that go to make us up, each one is an individual self-centred life is no mere phrase. It is not a mere convenience for descriptive purposes. The cell as a component of the body is not only a visibly demarcated unit but a unit-life centred on itself. It leads its own life . . . The cell is a unit-life, and our life which in its turn is a unitary life consists utterly of the cell-lives.[1]

But this story can be followed up in more detail and more concretely. Both the pathology of the brain and physiological investigations on sense perception speak unequivocally in favour of a regional separation of the sensorium into domains whose far-reaching independence is amazing because it would let us expect to find these regions associated with independent domains of the mind; but they are not. A particularly characteristic instance is the following. If you look at a distant landscape first in the ordinary way with both eyes open, then with the right eye alone, shutting the left, then the other way round, you find no noticeable difference. The psychic visional space is in all three cases identically the same. Now this might very well be due to the fact that from corresponding nerve-ends on the retina the stimulus is transferred to the same centre in the brain where 'the perception is manufactured' – just as, for example, in my house the knob at the entrance door and the one in my wife's bedroom activate the same bell, situated above the kitchen door. This would be the easiest explanation; but it is wrong.

Sherrington tells us of very interesting experiments on the threshold frequency of flickering. I shall try to give you as brief an account as possible. Think of a miniature lighthouse set up in the laboratory and giving off a great many flashes per second, say 40 or 60 or 80 or 100. As you increase the frequency of the flashes the flickering disappears at a definite frequency, depending on the experimental details; and the onlooker, whom we suppose to watch with both eyes in the ordinary way, sees then a continuous light.[2] Let this threshold frequency be 60 per second in given circumstances. Now in a

[1] *Man on his Nature*, 1st edn (1940), p. 73.
[2] In this way the fusion of successive pictures is produced in the cinema.

second experiment, with nothing else changed, a suitable contraption allows only every second flash to reach the right eye, every other flash to reach the left eye, so that every eye receives only 30 flashes per second. If the stimuli were conducted to the same physiological centre, this should make no difference: if I press the button before my entrance door, say every two seconds, and my wife does the same in her bedroom, but alternately with me, the kitchen bell will ring every second, just the same as if one of us had pressed his button every second or both of us had done so synchronously every second. However, in the second flicker experiment this is not so. Thirty flashes to the right eye plus alternating 30 flashes to the left are far from sufficient to remove the sensation of flickering; double the frequency is required for that, namely, 60 to the right and 60 to the left, if both eyes are open. Let me give you the main conclusion in Sherrington's own words:

It is not spatial conjunction of cerebral mechanism which combines the two reports . . . It is much as though the right- and left-eye images were seen each by one of two observers and the minds of the two observers were combined to a single mind. It is as though the right-eye and left-eye perceptions are elaborated singly and then psychically combined to one . . . It is as if each eye had a separate sensorium of considerable dignity proper to itself, in which mental processes based on that eye were developed up to even full perceptual levels. Such would amount physiologically to a visual sub-brain. There would be two such sub-brains, one for the right eye and one for the left eye. Contemporaneity of action rather than structural union seems to provide their mental collaboration.[1]

This is followed by very general considerations, of which I shall again pick out only the most characteristic passages:

Are there thus quasi-independent sub-brains based on the several modalities of sense? In the roof-brain the old 'five' senses instead of being merged inextricably in one another and further submerged under mechanism of higher order are still plain to find, each demarcated in its separate sphere. How far is the mind a collection of quasi-independent perceptual minds integrated psychically in large measure by temporal concurrence of experience? . . . When it

[1] *Man on his Nature*, pp. 273–5.

is a question of 'mind' the nervous system does not integrate itself by centralization upon a pontifical cell. Rather it elaborates a millionfold democracy whose each unit is a cell . . . the concrete life compounded of sublives reveals, although integrated, its additive nature and declares itself an affair of minute foci of life acting together . . . When however we turn to the mind there is nothing of all this. The single nerve-cell is never a miniature brain. The cellular constitution of the body need not be for any hint of it from 'mind' . . . A single pontifical brain-cell could not assure to the mental reaction a character more unified, and non-atomic than does the roof-brain's multitudinous sheet of cells. Matter and energy seem granular in structure, and so does 'life', but not so mind.

I have quoted you the passages which have most impressed me. Sherrington, with his superior knowledge of what is actually going on in a living body, is seen struggling with a paradox which in his candidness and absolute intellectual sincerity he does not try to hide away or explain away (as many others would have done, nay have done), but he almost brutally exposes it, knowing very well that this is the only way of driving any problem in science or philosophy nearer towards its solution, while by plastering it over with 'nice' phrases you prevent progress and make the antinomy perennial (not forever, but until someone notices your fraud). Sherrington's paradox too is an arithmetical paradox, a paradox of numbers, and it has, so I believe, very much to do with the one to which I had given this name earlier in this chapter, though it is by no means identical with it. The previous one was, briefly, the *one* world crystallizing out of the many minds. Sherrington's is the *one* mind, based ostensibly on the many cell-lives or, in another way, on the manifold sub-brains, each of which seems to have such a considerable dignity proper to itself that we feel impelled to associate a sub-mind with it. Yet we know that a sub-mind is an atrocious monstrosity, just as is a plural-mind – neither having any counterpart in anybody's experience, neither being in any way imaginable.

I submit that both paradoxes will be solved (I do not pretend to solve them here and now) by assimilating into our

Western build of science the Eastern doctrine of identity. Mind is by its very nature a *singulare tantum*. I should say: the over-all number of minds is just one. I venture to call it indestructible since it has a peculiar timetable, namely mind is always *now*. There is really no before and after for mind. There is only a now that includes memories and expectations. But I grant that our language is not adequate to express this, and I also grant, should anyone wish to state it, that I am now talking religion, not science – a religion, however, not opposed to science, but supported by what disinterested scientific research has brought to the fore.

Sherrington says: 'Man's mind is a recent product of our planet's side.'[1]

I agree, naturally. If the first word (man's) were left out, I would not. We dealt with this earlier, in chapter 1. It would seem queer, not to say ridiculous, to think that the contemplating, conscious mind that alone reflects the becoming of the world should have made its appearance only at some time in the course of this 'becoming', should have appeared contingently, associated with a very special biological contraption which in itself quite obviously discharges the task of facilitating certain forms of life in maintaining themselves, thus favouring their preservation and propagation: forms of life that were late-comers and have been preceded by many others that maintained themselves without that particular contraption (a brain). Only a small fraction of them (if you count by species) have embarked on 'getting themselves a brain'. And before that happened, should it all have been a performance to empty stalls? Nay, may we call a world that nobody contemplates even that? When an archaeologist reconstructs a city or a culture long bygone, he is interested in human life in the past, in actions, sensations, thoughts, feelings, in joy and sorrow of humans, displayed there and then. But a world existing for many millions of years without any mind being aware of it, contemplating it, is it anything at all? Has it existed? For do not let us forget: to say, as we did, that the becoming of the world is reflected in a conscious mind is but a

[1] *Man on his Nature*, p. 218.

cliché, a phrase, a metaphor that has become familiar to us. The world is given but once. Nothing is reflected. The original and the mirror-image are identical. The world extended in space and time is but our representation (*Vorstellung*). Experience does not give us the slightest clue of its being anything besides that – as Berkeley was well aware.

But the romance of a world that had existed for many millions of years before it, quite contingently, produced brains in which to look at itself has an almost tragic continuation that I should like to describe again in Sherrington's words:

The universe of energy is we are told running down. It tends fatally towards an equilibrium which shall be final. An equilibrium in which life cannot exist. Yet life is being evolved without pause. Our planet in its surround has evolved it and is evolving it. And with it evolves mind. If mind is not an energy-system how will the running down of the universe affect it? Can it go unscathed? Always so far as we know the finite mind is attached to a running energy-system. When that energy-system ceases to run what of the mind which runs with it? Will the universe which elaborated and is elaborating the finite mind then let it perish?[1]

Such considerations are in some way disconcerting. The thing that bewilders us is the curious double role that the conscious mind acquires. On the one hand it is the stage, and the only stage on which this whole world-process takes place, or the vessel or container that contains it all and outside which there is nothing. On the other hand we gather the impression, maybe the deceptive impression, that within this world-bustle the conscious mind is tied up with certain very particular organs (brains), which while doubtless the most interesting contraption in animal and plant physiology are yet not unique, not *sui generis*; for like so many others they serve after all only to maintain the lives of their owners, and it is only to this that they owe their having been elaborated in the process of speciation by natural selection.

Sometimes a painter introduces into his large picture, or a poet into his long poem, an unpretending subordinate character who is himself. Thus the poet of the *Odyssey* has, I suppose,

[1] *Man on his Nature*, p. 232.

meant himself by the blind bard who in the hall of the Phaea-
cians sings about the battles of Troy and moves the battered
hero to tears. In the same way we meet in the song of the
Nibelungs, when they traverse the Austrian lands, with a poet
who is suspected to be the author of the whole epic. In Dürer's
All-Saints picture two circles of believers are gathered in prayer
around the Trinity high up in the skies, a circle of the blessed
above, and a circle of humans on the earth. Among the latter
are kings and emperors and popes, but also, if I am not
mistaken, the portrait of the artist himself, as a humble side-
figure that might as well be missing.

To me this seems to be the best simile of the bewildering
double role of mind. On the one hand mind is the artist who has
produced the whole; in the accomplished work, however, it is
but an insignificant accessory that might be absent without
detracting from the total effect.

Speaking without metaphor we have to declare that we are
here faced with one of these typical antinomies caused by the
fact that we have not yet succeeded in elaborating a fairly
understandable outlook on the world without retiring our own
mind, the producer of the world picture, from it, so that mind
has no place in it. The attempt to press it into it, after all,
necessarily produces some absurdities.

Earlier I have commented on the fact that for this same
reason the physical world picture lacks all the sensual qualities
that go to make up the Subject of Cognizance. The model is
colourless and soundless and unpalpable. In the same way and
for the same reason the world of science lacks, or is deprived of,
everything that has a meaning only in relation to the con-
sciously contemplating, perceiving and feeling subject. I mean
in the first place the ethical and aesthetical values, any values of
any kind, everything related to the meaning and scope of the
whole display. All this is not only absent but it cannot, from the
purely scientific point of view, be inserted organically. If one
tries to put it in or on, as a child puts colour on his uncoloured
painting copies, it will not fit. For anything that is made to enter
this world model willy-nilly takes the form of scientific assertion
of facts; and as such it becomes wrong.

Life is valuable in itself. 'Be reverent towards life' is how Albert Schweitzer has framed the fundamental commandment of ethics. Nature has no reverence towards life. Nature treats life as though it were the most valueless thing in the world. Produced million-fold it is for the greatest part rapidly annihilated or cast as prey before other life to feed it. This precisely is the master-method of producing ever-new forms of life. 'Thou shalt not torture, thou shalt not inflict pain!' Nature is ignorant of this commandment. Its creatures depend upon racking each other in everlasting strife.

'There is nothing either good or bad but thinking makes it so.' No natural happening is in itself either good or bad, nor is it in itself either beautiful or ugly. The values are missing, and quite particularly meaning and end are missing. Nature does not act by purposes. If in German we speak of a purposeful (*zweckmässig*) adaptation of an organism to its environment, we know this to be only a convenient way of speech. If we take it literally, we are mistaken. We are mistaken within the frame of our world picture. In it there is only causal linkage.

Most painful is the absolute silence of all our scientific investigations towards our questions concerning the meaning and scope of the whole display. The more attentively we watch it, the more aimless and foolish it appears to be. The show that is going on obviously acquires a meaning only with regard to the mind that contemplates it. But what science tells us about this relationship is patently absurd: as if mind had only been produced by that very display that it is now watching and would pass away with it when the sun finally cools down and the earth has been turned into a desert of ice and snow.

Let me briefly mention the notorious atheism of science which comes, of course, under the same heading. Science has to suffer this reproach again and again, but unjustly so. No personal god can form part of a world model that has only become accessible at the cost of removing everything personal from it. We know, when God is experienced, this is an event as real as an immediate sense perception or as one's own personality. Like them he must be missing in the space-time picture. I do not find God anywhere in space and time – that is what the honest naturalist tells you. For this he incurs blame from him in whose catechism is written: God is spirit.

CHAPTER 5

Science and Religion

Can science vouchsafe information on matters of religion? Can the results of scientific research be of any help in gaining a reasonable and satisfactory attitude towards those burning questions which assail everyone at times? Some of us, in particular healthy and happy youth, succeed in shoving them aside for long periods; others, in advanced age, have satisfied themselves that there is no answer and have resigned themselves to giving up looking for one, while others again are haunted throughout their lives by this incongruity of our intellect, haunted also by serious fears raised by time-honoured popular superstition. I mean mainly the questions concerned with the 'other world', with 'life after death', and all that is connected with them. Notice please that I shall not, of course, attempt to answer *these* questions, but only the much more modest one, whether science can give any information about them or aid our – to many of us unavoidable – thinking about them.

To begin with, in a very primitive way it certainly can, and has done so without much ado. I remember seeing old prints, geographical maps of the world, so I believe, including hell, purgatory and heaven, the former being placed deep underground, the latter high above in the skies. Such representations were not meant purely allegorically (as they might be in later periods, for example, in Dürer's famous *All-Saints* picture); they testify to a crude belief quite popular at the time. Today no church requests the faithful to interpret its dogmas in this materialistic fashion, nay it would seriously discourage such an attitude. This advancement has certainly been aided by our knowledge of the interior of our planet (scanty though it be), of the nature of volcanoes, of the composition of our atmosphere, of the probable history of the solar system and of the structure of the galaxy and the universe. No cultured person would expect to find these dogmatic figments in any region of that part of space which is accessible to our investigation, I daresay not even in a region continuing that space but inaccessible to research; he would give them, even if convinced of their reality, a spiritual standing. I will not say that with deeply religious persons such enlightenment had to await the aforesaid findings of science, but they have certainly helped in eradicating materialistic superstition in those matters.

However, this refers to a rather primitive state of mind. There are points of greater interest. The most important contributions from science to overcome the baffling questions 'Who are we really? Where have I come from and where am I going?' – or at least to set our minds at rest – I say, the most appreciable help science has offered us in this is, in my view, the gradual idealization of time. In thinking of this the names of three men obtrude themselves upon us, though many others, including non-scientists, have hit on the same groove, such as St Augustine of Hippo and Boethius; the three are Plato, Kant and Einstein.

The first two were not scientists, but their keen devotion to philosophic questions, their absorbing interest in the world, originated from science. In Plato's case it came from mathematics and geometry (the 'and' would be out of place today, but not, I think, in his time). What has endowed Plato's life-work with such unsurpassed distinction that it shines in undiminished splendour after more than two thousand years? For all we can tell, no special discovery about numbers or geometrical figures is to his credit. His insight into the material world of physics and life is occasionally fantastic and altogether inferior to that of others (the sages from Thales to Democritus) who lived, some of them more than a century, before his time; in knowledge of nature he was widely surpassed by his pupil Aristotle and by Theophrastus. To all but his ardent worshippers long passages in his dialogues give the impression of a gratuitous quibbling on words, with no desire to define the meaning of a word, rather in the belief that the word itself will display its content if you turn it round and round long enough. His social and political Utopia, which failed and put him into grave danger when he tried to promote it practically, finds few admirers in our days, that have sadly experienced the like. So what made his fame?

In my opinion it was this, that he was the first to envisage the idea of timeless existence and to emphasize it – against reason – as a reality, more real than our actual experience; this, he said, is but a shadow of the former, from which all experienced reality is borrowed. I am speaking of the theory of forms (or ideas). How did it originate? There is no doubt that it was aroused by his becoming acquainted with the teaching of Parmenides and the Eleatics. But it is equally obvious that this met in Plato with an alive congenial vein, an occurrence very much on the line of Plato's own beautiful simile that learning by reason has the nature of remembering knowledge, previously possessed but at the time latent, rather than that of discovering entirely new verities. However, Parmenides' everlasting, ubiquitous and changeless One has in Plato's mind turned into a much more powerful thought, the Realm of Ideas, which appeals to the imagination, though, of necessity, it remains a mystery. But this thought sprang, as I believe, from a very real experience, namely, that he was struck with admiration and awe by the revelations in the realm of numbers and geometrical figures – as many a man was after him and the Pythagoreans were before. He recognized and absorbed deeply into his mind the nature of these revelations, that they unfold themselves by pure logical reasoning, which makes us acquainted with true relations whose truth is not only unassailable, but is obviously there, forever; the relations held and will hold irrespective of our inquiry into them. A mathematical truth is timeless, it does not come into being when we discover it. Yet its discovery is a very real event, it may be an emotion like a great gift from a fairy.

The three heights of a triangle (ABC) meet at one point (O).

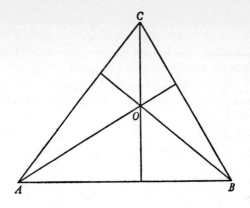

Fig. 1.

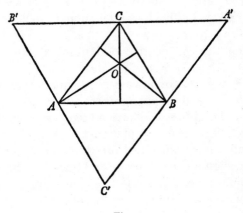

Fig. 2.

(Height is the perpendicular, dropped from a corner onto the side opposite to it, or onto its prolongation.) At first sight one does not see why they should; *any* three lines do not, they usually form a triangle. Now draw through every corner the parallel to the opposite side, to form the bigger triangle $A'B'C'$. It consists of four congruent triangles. The three heights of ABC are in the bigger triangle the perpendiculars erected in the middle of its sides, their 'symmetry lines'. Now the one erected at C must contain all the points that have the same distance from A' as from B'; the one erected at B contains all those points that have the same distance from A' as from C'. The point where these two perpendiculars meet has therefore the same distance from all three corners A', B', C', and must therefore lie also on the perpendicular erected at A because this one contains all points that have the same distance from B' as from C'. Q.E.D.

Every integer, except 1 and 2, is 'in the middle' of two prime numbers, or is their arithmetical mean; for instance

$$8 = \tfrac{1}{2}(5+11) = \tfrac{1}{2}(3+13)$$
$$17 = \tfrac{1}{2}(3+31) = \tfrac{1}{2}(29+5) = \tfrac{1}{2}(23+11)$$
$$20 = \tfrac{1}{2}(11+29) = \tfrac{1}{2}(3+37).$$

As you see, there is usually more than one solution. The theorem is called Goldbach's and is thought to be true, though it has not been proved.

By adding the consecutive odd numbers, thus first taking just 1, then $1+3 = 4$, then $1+3+5 = 9$, then $1+3+5+7 = 16$, you always get a square number, indeed you get in this way all square numbers, always the square of the number of odd numbers you have added. To grasp the generality of this relation one may replace in the sum the summands of every pair that is equidistant from the middle (thus: the first and the last, then the first but one and the last but one, etc.) by their arithmetic mean, which is obviously just equal to the number of summands; thus, in the last of the above examples:

$$4+4+4+4 = 4 \times 4.$$

Let us now turn to Kant. It has become a commonplace that he taught the ideality of space and time and that this was a fundamental, if not the most fundamental part of his teaching. Like most of it, it can be neither verified nor falsified, but it does not lose interest on this account (rather it gains; if it could be proved or disproved it would be trivial). The meaning is that, to be spread out in space and to happen in a well-defined temporal order of 'before and after' is not a quality of the world that we perceive, but pertains to the perceiving mind which, in its present situation anyhow, cannot help registering anything that is offered to it according to these two card-indexes, space and time. It does not mean that the mind comprehends these order-schemes irrespective of, and before, any experience, but that it cannot help developing them and applying them to experience when this comes along, and particularly that this fact does not prove or suggest space and time to be an order-scheme inherent in that 'thing-in-itself' which, as some believe, causes our experience.

It is not difficult to make a case that this is humbug. No single man can make a distinction between the realm of his perceptions and the realm of things that cause it since, however detailed the knowledge he may have acquired about the whole story, the story is occurring only once not twice. The duplication is an allegory, suggested mainly by communication with other human beings and even with animals; which shows that their perceptions in the same situation seem to be very similar to his own apart from insignificant differences in the point of view – in the literal meaning of 'point of projection'. But even supposing that this compels us to consider an objectively existing world the cause of our perceptions, as most people do, how on earth shall we decide that a common feature of all our experience is due to the constitution of our mind rather than a quality shared by all those objectively existing things? Admittedly our sense perceptions constitute our sole knowledge about things. This objective world remains a hypothesis, however natural. If we do adopt it, is it not by far the most natural thing to ascribe to that external world, and not to ourselves, all the characteristics that our sense perceptions find in it?

However, the supreme importance of Kant's statement does not consist in justly distributing the roles of the mind and its object – the world – between them in the process of 'mind forming an idea of the world', because, as I just pointed out, it is hardly possible to discriminate the two. The great thing was to form the idea that this *one thing* – mind or world – may well be capable of other forms of appearance that we cannot grasp and that do not imply the notions of space and time. This means an imposing liberation from our inveterate prejudice. There probably are other orders of appearance than the space-time-like. It was, so I believe, Schopenhauer who first read this from Kant. This liberation opens the way to belief, in the religious sense, without running all the time against the clear results which experience about the world as we know it and plain thought unmistakably pronounce. For instance – to speak of the most momentous example – experience as we know it unmistakably obtrudes the conviction that it cannot survive the destruction of the body, with whose life, as we know life, it is inseparably bound up. So is there to be nothing after this life? No. Not in the way of experience as we know it necessarily to take place in space and time. But, in an order of appearance in which time plays no part, this notion of 'after' is meaningless. Pure thinking cannot, of course, procure us a guarantee that there *is* that sort of thing, But it can remove the apparent obstacles to conceiving it as possible. That is what Kant has done by his analysis, and that, to my mind, is his philosophical importance.

I now come to speak about Einstein in the same context. Kant's attitude towards science was incredibly naïve, as you will agree if you turn the leaves of his *Metaphysical Foundations of Science* (*Metaphysische Anfangsgründe der Naturwissenschaft*). He accepted physical science in the form it had reached during his lifetime (1724–1804) as something more or less final and he busied himself to account for its statements philosophically. This happening to a great genius ought to be a warning to philosophers ever after. He would show plainly that space was necessarily infinite and believed firmly that it was in the nature of the human mind to endow it with the geometrical properties summarized by Euclid. In this Euclidean space a mollusc of matter moved, that is, changed its configuration as time went on. To Kant, as to any physicist of his period, space and time were two entirely different conceptions, so he had no qualms in calling the former the form of our external intuition, and time the form of our internal intuition (*Anschauung*). The recognition that Euclid's infinite space is not a necessary way of looking at the world of our experience and that space and time are better looked upon as one continuum of four dimensions seemed to shatter Kant's foundation – but actually did no harm to the more valuable part of his philosophy.

This recognition was left to Einstein (and several others, H. A. Lorentz, Poincaré, Minkowski, for example). The mighty impact of their discoveries on philosophers, men-in-the-street, and ladies in the drawing-room is due to the fact that they brought it to the fore: even in the domain of our experience the spatio-temporal relations are much more intricate than Kant dreamed them to be, following in this all previous physicists, men-in-the-street and ladies in the drawing-room.

The new view has its strongest impact on the previous notion of time. Time is the notion of 'before and after'. The new attitude springs from the following two roots:

(i) The notion of 'before and after' resides on the 'cause and effect' relation. We know, or at least we have formed the idea, that one event A can cause, or at least modify, another event B, so that if A were not, then B were not, at least not in this modified form. For instance when a shell explodes, it kills a man who was sitting on it; moreover the explosion is heard at distant places. The killing may be simultaneous to the explosion, the hearing of the sound at a distant place will be later; but certainly none of the effects can be earlier. This is a basic notion, indeed it is the one by which also in everyday life the question is decided which of two events was later or at least not earlier. The distinction rests entirely on the idea that the effect cannot precede the cause. If we have reasons to think that B has been caused by A, or that it at least shows vestiges of A, or even if (from some circumstantial evidence) it is conceivable that it shows vestiges, then B is deemed to be certainly not earlier than A.

(2) Keep this in mind. The second root is the experimental and observational evidence that effects do not spread with arbitrarily high velocity. There is an upper limit, which incidentally is the velocity of light in empty space. In human measure it is very high, it would go round the equator about

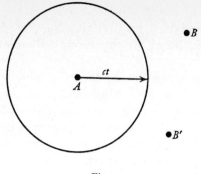

Fig. 3.

seven times in one second. Very high, but not infinite, call it c. Let this be agreed upon as a fundamental fact of nature. It then follows that the above-mentioned discrimination between 'before and after' or 'earlier and later' (based on the cause-and-effect relation) is not universally applicable, it breaks down in some cases. This is not as easily explained in non-mathematical language. Not that the mathematical scheme is so complicated. But everyday language is prejudicial in that it is so thoroughly imbued with the notion of time – you cannot use a verb (*verbum*, 'the' word, Germ. *Zeitwort*) without using it in one or the other tense.

The simplest but, as will turn out, not fully adequate consideration runs thus. Given an event A. Contemplate at any later time an event B outside the sphere of radius ct around A. Then B cannot exhibit any 'vestige' of A; nor, of course can A from B. Thus our criterion breaks down. By the language we used we have, of course, dubbed B to be the later. But are we right in this, since the criterion breaks down either way?

Contemplate at a time earlier (by t) an event B' outside that same sphere. In this case, just as before, no vestige of B' can have reached A (and, of course, none from A can be exhibited on B').

Thus in both cases there is exactly the same relationship of mutial non-interference. There is no conceptual difference between the classes B and B' with regard to their cause-effect relation to A. So if we want to make this relation, and not a linguistic prejudice, the basis of the 'before and after', then the B and B' form one class of events that are neither earlier nor later than A. The region of space-time occupied by this class is called the region of 'potential simultaneity' (with respect to event A). This expression is used, because a space-time frame can always be adopted that makes A simultaneous with a selected particular B or a particular B'. This was Einstein's discovery (which goes under the name of The Theory of Special Relativity, 1905).

Now these things have become very concrete reality to us physicists, we use them in everyday work just as we use the multiplication table or Pythagoras' theorem on right-angled triangles. I have sometimes wondered why they made such a great stir both among the general public and among philosophers. I suppose it is this, that it meant the dethronement of time as a rigid tyrant imposed on us from outside, a liberation from the unbreakable rule of 'before and after'. For indeed time is our most severe master by ostensibly restricting the existence of each of us to narrow limits – seventy or eighty years, as the Pentateuch has it. To be allowed to play about with such a master's programme believed unassailable until then, to play about with it albeit in a small way, seems to be a great relief, it seems to encourage the thought that the whole 'timetable' is probably not quite as serious as it appears at first sight. And this thought is a religious thought, nay I should call it *the* religious thought.

Einstein has not – as you sometimes hear – given the lie to Kant's deep thoughts on the idealization of space and time; he has, on the contrary, made a large step towards its accomplishment.

I have spoken of the impact of Plato, Kant and Einstein on the philosophical and religious outlook. Now between Kant and Einstein, about a generation before the latter, physical science had witnessed a momentous event which might have seemed calculated to stir the thoughts of philosophers, men-in-the-street and ladies in the drawing-room at least as much as the theory of relativity, if not more so. That this was not the case is, I believe, due to the fact that this turn of thought is even more difficult to understand and was therefore grasped by very few among the three categories of persons, at the best by one or another philosopher. This event is attached to the names of the American Willard Gibbs and the Austrian Ludwig Boltzmann. I will now say something about it.

With very few exceptions (that really are exceptions) the course of events in nature is irreversible. If we try to imagine a time-sequence of phenomena exactly opposite to one that is actually observed – as in a cinema film projected in reversed order – such a reversed sequence, though it can easily be imagined, would nearly always be in gross contradiction to well-established laws of physical science.

The general 'directedness' of all happening was explained by the mechanical or statistical theory of heat, and this explanation was duly hailed as its most admirable achievement. I cannot enter here on the details of the physical theory, and this is not necessary for grasping the gist of the explanation. This would have been very poor, had irreversibility been stuck in as a fundamental property of the microscopic mechanism of atoms and molecules. This would not have been better than many a medieval purely verbal explanation such as: fire is hot on account of its fiery quality. No. According to Boltzmann we are faced with the natural tendency of any state of order to turn on its own into a less orderly state, but not the other way round. Take as a simile a set of playing cards that you have carefully arranged, beginning with 7, 8, 9, 10, knave, queen, king, ace of hearts, then the same in diamonds, etc. If this well-ordered set is shuffled once, twice or three times it will gradually turn into a random set. But this is not an intrinsic property of the process of shuffling. Given the resulting disorderly set, a process of shuffling is perfectly thinkable that would exactly cancel the effect of the first shuffling and restore the original order. Yet everybody will expect the first course to take place, nobody the second – indeed he might have to wait pretty long for it to happen by chance.

Now this is the gist of Boltzmann's explanation of the unidirectional character of everything that happens in nature (including, of course, the life-history of an organism from birth to death). Its very virtue is that the 'arrow of time' (as Eddington called it) is not worked into the mechanisms of interaction, represented in our simile by the mechanical act of shuffling. This act, this mechanism is as yet innocent of any notion of past and future, it is in itself completely reversible, the 'arrow' – the very notion of past and future – results from statistical considerations. In our simile with the cards the point is this, that there is only one, or a very few, well-ordered arrangements of the cards, but billions of billions of disorderly ones.

Yet the theory has been opposed, again and again, occasionally by very clever people. The opposition boils down to this: the theory is said to be unsound on logical grounds. For, so it is said, if the basic mechanisms do not distinguish between the two directions of time, but work perfectly symmetrically in this respect, how should there from their co-operation result a behaviour of the whole, an integrated behaviour, that is strongly biased in one direction? Whatever holds for this direction must hold equally well for the opposite one.

If this argument is sound, it seems to be fatal. For it is aimed at the very point which was regarded as the chief virtue of the theory: to derive irreversible events from reversible basic mechanisms.

The argument is perfectly sound, yet it is not fatal. The argument is sound in asserting that what holds for one direction also holds for the opposite direction of time, which from the outset is introduced as a perfectly symmetrical variable. But you must not jump to the conclusion that it holds quite in general for both directions. In the most cautious wording one has to say that in any particular case it holds for either the one or the other direction. To this one must add: in the particular case of the world as we know it, the 'running down' (to use a phrase that has been occasionally adopted) takes place in one direction and this we call the direction from past to future. In other words the statistical theory of heat must be allowed to decide by itself high-handedly, by its own definition, in which direction time flows. (This has a momentous consequence for the methodology of the physicist. He must never introduce anything that decides independently upon the arrow of time, else Boltzmann's beautiful building collapses.)

It might be feared that in different physical systems the statistical definition of time might not always result in the same time-direction. Boltzmann boldly faced this eventuality; he maintained that if the universe is sufficiently extended and/or exists for a sufficiently long period, time might actually run in the opposite direction in distant parts of the world. The point has been argued, but it is hardly worth while arguing any longer. Boltzmann did not know what to us is at least extremely likely, namely that the universe, as we know it, is neither large enough nor old enough to give rise to such reversions on a large scale. I beg to be allowed to add without detailed explanations that on a very small scale, both in space and in time, such reversions have been observed (Brownian movement, Smoluchowski).

To my view the 'statistical theory of time' has an even stronger bearing on the philosophy of time than the theory of relativity. The latter, however revolutionary, leaves untouched the undirectional flow of time, which it presupposes, while the statistical theory constructs it from the order of the events. This means a liberation from the tyranny of old Chronos. What we in our minds construct ourselves cannot, so I feel, have dictatorial power over our mind, neither the power of bringing it to the fore nor the power of annihilating it. But some of you, I am sure, will call this mysticism. So with all due acknowledgment to the fact that physical theory is at all times relative, in that it depends on certain basic assumptions, we may, or so I believe, assert that physical theory in its present stage strongly suggests the indestructibility of Mind by Time.

CHAPTER 6

The Mystery of the Sensual Qualities

In this last chapter I wish to demonstrate in a little more detail the very strange state of affairs already noticed in a famous fragment of Democritus of Abdera – the strange fact that on the one hand all our knowledge about the world around us, both that gained in everyday life and that revealed by the most carefully planned and painstaking laboratory experiments, rests entirely on immediate sense perception, while on the other hand this knowledge fails to reveal the relations of the sense perceptions to the outside world, so that in the picture or model we form of the outside world, guided by our scientific discoveries, all sensual qualities are absent. While the first part of this statement is, so I believe, easily granted by everybody, the second half is perhaps not so frequently realized, simply because the non-scientist has, as a rule, a great reverence for science and credits us scientists with being able, by our 'fabulously refined methods', to make out what, by its very nature, no human can possibly make out and never will be able to make out.

If you ask a physicist what is his idea of yellow light, he will tell you that it is transversal electro-magnetic waves of wave-length in the neighbourhood of 590 millimicrons. If you ask him: But where does yellow come in? he will say: In my picture not at all, but these kinds of vibrations, when they hit the retina of a healthy eye, give the person whose eye it is the sensation of yellow. On further inquiry you may hear that different wave-lengths produce different colour-sensations, but not all do so, only those between about 800 and 400 $\mu\mu$. To the physicist the infra-red (more than 800 $\mu\mu$) and the ultra-violet (less than 400 μμ) waves are much the same kind of phenomena as those in the region between 800 and 400 μμ, to which the eye is sensitive. How does this peculiar selection come about? It is obviously an adaptation to the sun's radiation, which is strongest in this region of wave-lengths but falls off at either end. Moreover, the intrinsically brightest colour-sensation, the yellow, is encountered at that place (within the said region) where the sun's radiation exhibits its maximum, a true peak.

We may further ask: Is radiation in the neighbourhood of wave-length 590 μμ the only one to produce the sensation of yellow? The answer is: Not at all. If waves of 760 μμ, which by themselves produce the sensation of red, are mixed in a definite proportion with waves of 535 μμ, which by themselves produce the sensation of green, this mixture produces a yellow that is indistinguishable from the one produced by 590 μμ. Two adjacent fields illuminated, one by the mixture, the other by the single spectral light, look exactly alike, you cannot tell which is which. Could this be foretold from the wave-lengths – is there a numerical connection with these physical, objective characteristics of the waves? No. Of course, the chart of all mixtures of this kind has been plotted empirically; it is called the colour triangle. But it is not simply connected with the wave-lengths. There is no general rule that a mixture of two spectral lights matches one between them; for example a mixture of 'red' and 'blue' from the extremities of the spectrum gives 'purple', which is not produced by any single spectral light. Moreover, the said chart, the colour triangle, varies slightly from one person to the other, and differs considerably for some persons, called anomalous trichromates (who are *not* colour-blind).

The sensation of colour cannot be accounted for by the physicist's objective picture of light-waves. Could the physiologist account for it, if he had fuller knowledge than he has of the processes in the retina and the nervous processes set up by them in the optical nerve bundles and in the brain? I do not think so. We could at best attain to an objective knowledge of what nerve fibres are excited and in what proportion, perhaps even to know exactly the processes they produce in certain brain cells – whenever your mind registers the sensation of yellow in a particular direction or domain of our field of vision. But even such intimate knowledge would not tell us anything about the sensation of colour, more particularly of yellow in this direction – the same physiological processes might conceivably result in a sensation of sweet taste, or anything else. I mean to say simply this, that we may be sure there is no nervous process whose objective description includes the characteristic 'yellow colour' or 'sweet taste', just as little as the objective description of an electro-magnetic wave includes either of these characteristics.

The same holds for other sensations. It is quite interesting to compare the perception of colour, which we have just surveyed, with that of sound. It is normally conveyed to us by elastic waves of compression and dilatation, propagated in the air. Their wave-length – or to be more accurate their frequency – determines the pitch of the sound heard. (N.B. The physiological relevance pertains to the frequency, not to the wave-length, also in the case of light, where, however, the two are virtually exact reciprocals of each other, since the velocities of propagation in empty space and in air do not differ perceptibly.) I need not tell you that the range of frequencies of 'audible sound' is very different from that of 'visible light', it ranges from about 12 or 16 per second to 20,000 or 30,000 per second, while those for light are of the order of several hundred (English) billions. The relative range, however, is much wider for sound, it embraces about 10 octaves (against hardly one for 'visible light'); moreover, it changes with the individual, especially with age: the upper limit is regularly and considerably reduced as age advances. But the most striking fact about sound is that a mixture of several distinct frequencies never combines to produce just one intermediate pitch such as could be produced by one intermediate frequency. To a large extent the superposed pitches are perceived separately – though simultaneously – especially by highly musical persons. The admixture of many higher notes ('overtones') of various qualities and intensities results in

what is called the timbre (German: *Klangfarbe*), by which we
learn to distinguish a violin, a bugle, a church bell, piano . . .
even from a single note that is sounded. But even noises have
their timbre, from which we may infer what is going on; and
even my dog is familiar with the peculiar noise of the opening
of a certain tin box, out of which he occasionally receives a
biscuit. In all this the ratios of the co-operating frequencies
are all-important. If they are all changed in the same ratio, as
on playing a gramophone record too slow or too fast, you still
recognize what is going on. Yet some relevant distinctions
depend on the absolute frequencies of certain components. If a
gramophone record containing a human voice is played too
fast, the vowels change perceptibly, in particular the 'a' as in
'car' changes into that in 'care'. A continuous range of
frequencies is always disagreeable, whether offered as a
sequence, as by a siren or a howling cat, or simultaneously,
which is difficult to implement, except perhaps by a host of
sirens or a regiment of howling cats. This is again entirely
different from the case of light perception. All the colours
which we normally perceive are produced by continuous
mixtures; and a continuous gradation of tints, in a painting or
in nature, is sometimes of great beauty.

The chief characteristics of sound perception are well
understood in the mechanism of the ear, of which we have
better and safer knowledge than of the chemistry of the retina.
The principal organ is the *cochlea*, a coiled bony tube which
resembles the shell of a certain type of sea-snail: a tiny
winding staircase that gets narrower and narrower as it
'ascends'. In place of the steps (to continue our simile), across
the winding staircase elastic fibres are stretched, forming a
membrane, the width of the membrane (or the length of the
individual fibre) diminishing from the 'bottom' to the 'top'.
Thus, like the strings of a harp or a piano, the fibres of
different length respond mechanically to oscillations of different frequency. To a definite frequency a definite small area of
the membrane – not just one fibre – responds, to a higher
frequency another area, where the fibres are shorter. A
mechanical vibration of definite frequency must set up, in

each of that group of nerve fibres, the well-known nerve impulses that are propagated to certain regions of the cerebral cortex. We have the general knowledge that the process of conduction is very much the same in all nerves and changes only with the intensity of excitation; the latter affects the frequency of the pulses, which, of course, must not be confused with the frequency of sound in our case (the two have nothing to do with each other).

The picture is not as simple as we might wish it to be. Had a physicist constructed the ear, with a view to procuring for its owner the incredibly fine discrimination of pitch and timbre that he actually possesses, the physicist would have constructed it differently. But perhaps he would have come back to it. It would be simpler and nicer if we could say that every single 'string' across the cochlea answers only to one sharply defined frequency of the incoming vibration. This is not so. But why is it not so? Because the vibrations of these 'strings' are strongly damped. This, of necessity, broadens their range of resonance. Our physicist might have constructed them with as little damping as he could manage. But this would have the terrible consequence that the perception of a sound would not cease almost immediately when the producing wave ceases; it would last for some time, until the poorly damped resonator in the cochlea died down. The discrimination of pitch would be obtained by sacrificing the discrimination in time between subsequent sounds. It is puzzling how the actual mechanism manages to reconcile both in a most consummate fashion.

I have gone into some detail here, in order to make you feel that neither the physicist's description, nor that of the physiologist, contains any trait of the sensation of sound. Any description of this kind is bound to end with a sentence like: those nerve impulses are conducted to a certain portion of the brain, where they are registered as a sequence of sounds. We can follow the pressure changes in the air as they produce vibrations of the ear-drum, we can see how its motion is transferred by a chain of tiny bones to another membrane and eventually to parts of the membrane inside the cochlea, composed of fibres of varying length, described above. We may reach an understanding of how such a vibrating fibre sets up an electrical and chemical process of conduction in the nervous fibre with which it is in touch. We may follow this conduction to the cerebral cortex and we may even obtain some objective knowledge of some of the things that happen there. But nowhere shall we hit on this 'registering as sound', which simply is not contained in our scientific picture, but is only in the mind of the person whose ear and brain we are speaking of.

We could discuss in similar manner the sensations of touch, of hot and cold, of smell and of taste. The latter two, the chemical senses as they are sometimes called (smell affording an examination of gaseous stuffs, taste that of fluids), have this in common with the visual sensation, that to an infinite number of possible stimuli they respond with a restricted manifold of sensate qualities, in the case of taste: bitter, sweet, sour and salty and their peculiar mixtures. Smell is, I believe, more various than taste, and particularly in certain animals it is much more refined than in man. What objective features of a physical or chemical stimulus modify the sensation notice-ably seems to vary greatly in the animal kingdom. Bees, for instance, have a colour vision reaching well into the ultra-violet; they are true trichromates (not dichromates, as they seemed in earlier experiments which paid no attention to the ultra-violet). It is of very particular interest that bees, as von Frisch in Munich found out not long ago, are peculiarly sensitive to traces of polarization of light; this aids their orientation with respect to the sun in a puzzlingly elaborate way. To a human being even completely polarized light is indistinguishable from ordinary, non-polarized light. Bats have been discovered to be sensible to extremely high fre-quency vibrations ('ultra-sound') far beyond the upper limit of human audition; they produce it themselves, using it as a sort of 'radar', to avoid obstacles. The human sense of hot or cold exhibits the queer feature of 'les extrêmes se touchent': if we inadvertently touch a very cold object, we may for a moment believe that it is hot and has burnt our fingers.

Some twenty or thirty years ago chemists in the U.S.A.

discovered a curious compound, of which I have forgotten the chemical name, a white powder, that is tasteless to some persons, but intensely bitter to others. This fact has aroused keen interest and has been widely investigated since. The quality of being a 'taster' (for this particular substance) is inherent in the individual, irrespective of any other conditions. Moreover, it is inherited according to the Mendel laws in a way familiar from the inheritance of blood group characteristics. Just as with the latter, there appears to be no conceivable advantage or disadvantage implied by your being a 'taster' or a 'non-taster'. One of the two 'alleles' is dominant in heterozygotes, I believe it is that of the taster. It seems to me very improbable that this substance, discovered haphazardly, should be unique. Very probably 'tastes differ' in quite a general way, and in a very real sense!

Let us now return to the case of light and probe a little deeper into the way it is produced and into the fashion in which the physicist makes out its objective characteristics. I suppose that by now it is common knowledge that light is usually produced by electrons, in particular by those in an atom where they 'do something' around the nucleus. An electron is neither red nor blue nor any other colour; the same holds for the proton, the nucleus of the hydrogen atom. But the union of the two in the atom of hydrogen, according to the physicist, produces electro-magnetic radiation of a certain discrete array of wave-lengths. The homogeneous constituents of this radiation, when separated by a prism or an optical grating, stimulate in an observer the sensations of red, green, blue, violet by the intermediary of certain physiological processes, whose general character is sufficiently well known to assert that they are not red or green or blue, in fact that the nervous elements in question display no colour in virtue of their being stimulated; the white or grey the nerve cells exhibit whether stimulated or not is certainly insignificant in respect of the colour sensation which, in the individual whose nerves they are, accompanies their excitation.

Yet our knowledge of the radiation of the hydrogen atom and of the objective, physical properties of this radiation originated from someone's observing those coloured spectral lines in certain positions within the spectrum obtained from glowing hydrogen vapour. This procured the first knowledge, but by no means the complete knowledge. To achieve it, the elimination of the sensates has to set in at once, and is worth pursuing in this characteristic example. The colour in itself tells you nothing about the wave-length; in fact we have seen before that, for example, a yellow spectral line might conceivably be not 'monochromatic' in the physicist's sense, but composed of many different wave-lengths, if we did not know that the construction of our spectroscope excludes this. It gathers light of a definite wave-length at a definite position in the spectrum. The light appearing there has always exactly the same colour from whatever source it stems. Even so the quality of the colour sensation gives no direct clue whatsoever to infer the physical property, the wave-length, and that quite apart from the comparative poorness of our discrimination of hues, which would not satisfy the physicist. *A priori* the sensation of blue might conceivably be stimulated by long waves and that of red by short waves, instead of the other way round, as it is.

To complete our knowledge of the physical properties of the light coming from any source a special kind of spectroscope has to be used; the decomposition is achieved by a diffraction grating. A prism would not do, because you do not know beforehand the angles under which it refracts the different wave-lengths. They are different for prisms of different material. In fact, *a priori*, with a prism you could not even tell that the more strongly deviated radiation is of shorter wave-length, as is actually the case.

The theory of the diffraction grating is much simpler than that of a prism. From the basic physical assumption about light – merely that it is a wave phenomenon – you can, if you have measured the number of the equidistant furrows of the grating per inch (usually of the order of many thousands), tell the exact angle of deviation for a given wave-length, and therefore, inversely, you can infer the wave-length from the 'grating constant' and the angle of deviation. In some cases (notably in the Zeeman and Stark effects) some of the spectral lines are polarized. To complete the physical description in this respect, in which the human eye is entirely insensitive, you put a polarizer (a Nicol prism) in the path of the beam, before decomposing it; on slowly rotating the Nicol around its axis certain lines are extinguished or reduced to minimal brightness for certain orientations of the Nicol, which indicate the direction (orthogonal to the beam) of their total or partial polarization.

Once this whole technique is developed, it can be extended far beyond the visible region. The spectral lines of glowing vapours are by no means restricted to the visible region, which is not distinguished physically. The lines form long, theoretically infinite series. The wave-lengths of each series are connected by a relatively simple mathematical law, peculiar to it, that holds uniformly throughout the series with no distinction of that part of the series that happens to lie in the visible region. These serial laws were first found empirically, but are now understood theoretically. Naturally, outside the visible region a photographic plate has to replace the eye. The wave-lengths are inferred from pure measurements of lengths: first, once and for all, of the grating constant, that is the distance between neighbouring furrows (the reciprocal of the number of furrows per unit length), then by measuring the positions of the lines on the photographic plate, from which, together with the known dimensions of the apparatus, the angles of deviation can be computed.

These are well-known things, but I wish to stress two points of general importance, which apply to well-nigh every physical measurement.

The state of affairs on which I have enlarged here at some length is often described by saying that, as the technique of measuring is refined, the observer is gradually replaced by more and more elaborate apparatus. Now this is, certainly in the present case, not true; he is not gradually replaced, but is so from the outset. I tried to explain that the observer's colourful impression of the phenomenon vouchsafes not the slightest clue to its physical nature. The device of ruling a grating and measuring certain lengths and angles has to be introduced, before even the roughest qualitative knowledge of what we call the objective physical nature of the light and of its physical components can be obtained. And this is the relevant step. That the device is later on gradually refined, while remaining essentially always the same, is epistemologically unimportant, however great the improvement achieved.

The second point is that the observer is never entirely replaced by instruments; for if he were, he could obviously obtain no knowledge whatsoever. He must have constructed the instrument and, either while constructing it or after, he must have made careful measurements of its dimensions and checks on its moving parts (say a supporting arm turning around a conical pin and sliding along a circular scale of angles) in order to ascertain that the movement is exactly the intended one. True, for some of these measurements and check-ups the physicist will depend on the factory that has produced and delivered the instrument; still all this information goes back ultimately to the sense perceptions of some living person or persons, however many ingenious devices may have been used to facilitate the labour. *Finally* the observer must, in using the instrument for his investigation, take readings on it, be they direct readings of angles or of distances, measured under the microscope, or between spectral lines recorded on a photographic plate. Many helpful devices can facilitate this work, for instance photometric recording across the plate of its transparency, which yields a magnified diagram on which the positions of the lines can be easily read. But they must be read! The observer's senses have to step in eventually. The most careful record, when not inspected, tells us nothing.

So we come back to this strange state of affairs. While the direct sensual perception of the phenomenon tells us nothing as to its objective physical nature (or what we usually call so) and has to be discarded from the outset as a source of information, yet the theoretical picture we obtain eventually rests entirely on a complicated array of various informations, all obtained by direct sensual perception. It resides upon them, it is pieced together from them, yet it cannot really be said to contain them. In using the picture we usually forget about them, except in the quite general way that we know our idea of a light-wave is not a haphazard invention of a crank but is based on experiment.

I was surprised when I discovered for myself that this state of affairs was clearly understood by the great Democritus in the fifth century B.C., who had no knowledge of any physical measuring devices remotely comparable to those I have been telling you about (which are of the simplest used in our time).

Galenus has preserved us a fragment (Diels, fr. 125), in which Democritus introduces the intellect (διάνοια) having an argument with the senses (αἰσθήσεις) about what is 'real'. The former says: 'Ostensibly there is colour, ostensibly sweetness, ostensibly bitterness, actually only atoms and the void', to which the senses retort: 'Poor intellect, do you hope to defeat us while from us you borrow your evidence? Your victory is your defeat.'

In this chapter I have tried by simple examples, taken from the humblest of sciences, namely physics, to contrast the two general facts (*a*) that all scientific knowledge is based on sense perception, and (*b*) that none the less the scientific views of natural processes formed in this way lack all sensual qualities and therefore cannot account for the latter. Let me conclude with a general remark.

Scientific theories serve to facilitate the survey of our observations and experimental findings. Every scientist knows how difficult it is to remember a moderately extended group of facts, before at least some primitive theoretical picture about them has been shaped. It is therefore small wonder, and by no means to be blamed on the authors of original papers or of text-books, that after a reasonably coherent theory has been formed, they do not describe the bare facts they have found or wish to convey to the reader, but clothe them in the terminology of that theory or theories. This procedure, while very useful for our remembering the facts in a well-ordered pattern,

tends to obliterate the distinction between the actual observations and the theory arisen from them. And since the former
always are of some sensual quality, theories are easily thought
to account for sensual qualities; which, of course, they never
do.

AUTOBIOGRAPHICAL SKETCHES

I lived far apart from my best friend, actually the only close friend I ever had, for the greater part of my life. (Maybe that is why I have often been accused of flirtatiousness instead of true friendship.) He studied biology (botany to be exact); I physics. And many a night we would stroll back and forth between Gluckgasse and Schlüsselgasse engrossed in philosophical conversation. Little did we know then that what seemed original to us had occupied great minds for centuries already. Don't teachers always do their best to avoid these topics for fear that they might conflict with religious doctrines and cause uncomfortable questions? This is the main reason for my turning against religion, which has never done me any harm.

I am not sure whether it was right after the First World War or during the time I spent in Zurich (1921–7) or even later in Berlin (1927–33) that Fränzel and I spent a long evening together again. The small hours of the morning found us still talking in a café on the outskirts of Vienna. He seemed to have changed a lot with the years. After all, our letters had been few and far between and of very little substance.

I might have added earlier that we also spent our time together reading Richard Semon. Never before or after did I read a serious book with anyone else. Richard Semon was soon banned by the biologists, since his views, as they saw them, were based on the inheritance of acquired characteristics. So his name was forgotten. Many years later I encountered him in a book (*Human Knowledge?*) by Bertrand Russell, who devoted a thorough study to this genial biologist, stressing the significance of his Mneme theory.

Fränzel and I did not see each other again until 1956. This time it was a very brief encounter in our flat in Vienna, Pasteurgasse 4, while others were present, so that those fifteen minutes are hardly worth mentioning. Fränzel and his wife lived across the border, our northern one, unhampered by the authorities, it seemed; nevertheless, leaving the country had become rather difficult. We never met again: two years later he died very suddenly.

Today I am still friends with his charming nephew and

niece, his favourite brother Silvio's children. Silvio, the youngest in the family, was a doctor in Krems, where I went to see him when I returned to Austria in 1956. He must have been seriously ill already, for he died not long afterwards. One of Fränzel's brothers, E., is still alive. He is a respected surgeon in Klagenfurt. E. once took me up the Einser (Sextener Dolomites) and, what's more, saw me safely down again. I am afraid we have lost contact, driven apart by our different views of the world.

Shortly before I entered the University of Vienna in 1906, the only university I was ever enrolled in, the great Ludwig Boltzmann met his sad end in Duino. To this day I have not forgotten the clear, precise and yet still enthusiastic words with which Fritz Hasenöhrl described Boltzmann's work to us. Boltzmann's scholar and successor held his inaugural address in autumn 1907 in the primitive lecture hall of the old Türkenstrasse building without any pomp or ceremony. I was deeply impressed by his introduction, and no perception in physics has ever seemed more important to me than that of Boltzmann – despite Planck and Einstein. Incidentally, Einstein's early work (before 1905) shows how fascinated he too was by Boltzmann's work. He was the only one who took a major step beyond it by inverting Boltzmann's equation $S = k \lg W$. No other human being had a greater influence on me than Fritz Hasenöhrl – except perhaps my father Rudolph, who in the course of those many years we lived together drew me into conversations concerning his many interests. But more about that later.

While still a student I made friends with Hans Thirring. This turned out to be a lasting relationship. When Hasenöhrl was killed in action in 1916, Hans Thirring became his successor; he retired at seventy, forgoing the privilege of remaining for the honorary year and leaving Boltzmann's professorial chair to his son, Walter.

After 1911, while I was assistant to Exner, I met K. W. F. Kohlrausch, and yet another lasting friendship began. Kohlrausch had made his name by proving experimentally the existence of the so-called 'Schweidle Fluctuations'. In the year before the outbreak of the war we worked together on the research of 'secondary radiations', which produced – at the smallest possible angle on small plates of varying material – a (mixed) beam of gamma rays. I learnt two things in those years: firstly that I was not suited to experimental work, and secondly that my surroundings and the people who were part of them were no longer capable of making experimental progress on a big scale. There were many reasons for this, one of them being that in charming old Vienna well-meaning blunderers were placed, often according to seniority, in key positions, thus impeding all progress. If only it had been realized that personalities with great mental capacities were needed, even if it meant bringing them in from afar! The theories of atmospheric electricity and radio activity were both originally developed in Vienna, but anyone who felt really dedicated to their work had to follow those theories wherever they had been passed on. Lise Meitner, for instance, left Vienna and went to Berlin.

But back to myself: in retrospect I am very grateful that because of my reserve officer's training in 1910/11 I was appointed assistant to Fritz Exner and not to Hasenöhrl. It meant that I was able to experiment with K. W. F. Kohlrausch and make use of a number of beautiful instruments, take them to my room, especially the optical ones, and dabble with them to my heart's content. Thus I could set the interferometer, admire the spectra, mix colours, etc. This was also how I discovered – through the Rayleigh equation – the deuter anomaly of my eyes. Moreover I was committed to do the long practical course, so that I learnt to appreciate the significance of measuring. I wish there were more theoretical physicists who did.

In 1918 we had a kind of revolution. The Emperor Karl abdicated and Austria became a republic. Our everyday life remained much the same. However, my life was affected by the breaking up of the Empire. I had accepted a post as a lecturer in theoretical physics in Czernowitz and had already envisaged spending all my free time acquiring a deeper knowledge of philosophy, having just discovered Schopenhauer, who introduced me to the Unified Theory of the Upanishads.

For us Viennese the war and its consequences meant that we could no longer satisfy our basic needs. Hunger was the punishment the victorious Entente had chosen in retaliation for the unlimited U-boat war of their enemies, a war so atrocious that Prince Bismarck's heir and his followers could only outdo it in quantity, and not in quality, in the Second World War. Hunger prevailed throughout the country except on the farms, where our poor women were sent to ask for eggs, butter and milk. Despite the goods with which they paid – knitted garments, pretty petticoats, etc. – they were sneered at and treated like beggars.

In Vienna it had become virtually impossible to socialize and entertain friends. There was simply nothing to offer, and even the simplest dishes were reserved for Sunday lunch. In some ways this lack of social activities was compensated by the daily visit to the community kitchens. The *Gemeinschaftsküchen* were often referred to as *Gemeinheitsküchen* (*Gemeinschaft* = 'community'; *Gemeinheit* = 'a mean trick'). There we met for lunch. We had to be grateful to the women who considered it their responsibility to create meals out of nothing. It is no doubt easier to do this for 30 or 50 people than for three. Besides, relieving others of a burden must in itself be rewarding.

My parents and I met a number of people with similar interests there and some of them, the Radons, for example, both of them mathematicians, became great friends of our family.

I believe that in one way my parents and I were particularly disadvantaged. At that time we lived in a large flat (actually two flats made into one) on the fifth floor of a rather valuable building in the city, which belonged to my mother's father. It had no electric light, partly because my grandfather did not want to pay for having it installed and also because my father, in particular, had become so used to the excellent gas light at a time when light bulbs were still very expensive and inefficient that we really saw no need for them. And we had the old tiled stoves removed and replaced by solid gas stoves with copper reflectors – servants were hard to come by in those days, and we had hoped to make things easier for ourselves. Gas was also used for cooking, although we did still have an enormous old wood-burning stove standing in the kitchen. This was all very well until one day one of the higher bureaucratic offices, probably the city council, decreed that gas was to be rationed. From that day on every household was allowed one cubic metre per day regardless of how the fuel had to be used. If anyone was found using more, they were simply cut off.

In the summer of 1919 we went to Millstadt, Carinthia, and my father, who was sixty-two, showed the first signs of ageing and of what was to be his final illness, a fact we did not become aware of at the time. Whenever we went for a walk he would lag behind, especially where it got steep, and he would feign botanical curiosity to mask his exhaustion. From about 1902 on Father's main interest was botany. During the summer months he collected material for his studies, not for setting up a herbarium of his own, but for experimenting with his microscope and microtome. He had become a morphogeneticist and phylogeneticist and had abandoned his dedication to Italy's great painters and also his own artistic interests, which consisted of sketching innumerable landscapes. Father's rather bored reaction to our coaxing: 'Oh, Rudolph, do come on' and 'Mr Schrödinger, it's getting rather late', did not alarm us either; we were actually used to that; so we put it down to his absorbed concentration.

After our return to Vienna the signs became more apparent, but still we did not take them seriously as a warning: frequent and heavy bleeding from his nose and retina, and finally fluid in his legs. I think he knew long before everyone else that his end was near. Unfortunately this was just the time of the gas calamity mentioned above. We acquired carbon lamps, and he insisted on tending them himself. A dreadful stench spread from his beautiful library, which he had turned into a carbide laboratory. Twenty years earlier, when he had learnt to etch with Schmutzer, he had used the room to soak his copper and zinc plates in acids and chlorinated water; I was still at school then, and had shown great interest in his activities. But now I left him to his own devices. I was glad to be back at my beloved physics institute after serving in the war for almost four years. Besides, in autumn 1919 I became engaged to the girl who has been my wife for forty years now. I do not know whether my father had adequate medical treatment, but what I do know is that I should have looked after him better. I should have asked Richard von Wettstein, who was after all a good friend of his, to seek help at the medical faculty. Would better advice have slowed down his arteriosclerosis? And if so, would it have been to the advantage of a sick man? Only Father was fully aware of our financial situation after the closing down of our oilcloth and linoleum store on the Stephansplatz in 1917 (due to lack of stock).

He died peacefully on Christmas Eve 1919, in his old armchair.

The following year was that of rampant inflation, which meant the depreciation of Father's meagre bank account, which would never have kept my parents' heads above water anyway. The proceeds of the Persian rugs he had sold (with my consent!) dissolved into nothing; gone for ever were the microscopes, the microtome and a good part of his library, which I gave away for a song after his death. His greatest worry during the last months had been that at the ripe old age of thirty-two I was earning virtually nothing – 1,000 Austrian kronen (before tax, that is, for I am sure he listed it in his tax declaration except when I was an officer during the war). The only success of his son that he lived to see was that I had been offered (and had also accepted) a better-paid post as private lecturer and assistant to Max Wien in Jena.

My wife and I moved to Jena in April 1920, leaving my mother to fend for herself, in fact which I am not at all proud of today. She had to bear the burden of packing and clearing the flat. Oh, how blind we all were! Her father, who owned the house, was rather worried after my father's death about who would pay the rent. We were in no position to do so, and Mother had to make room for a more affluent tenant. My future father-in-law kindly turned up with the man, a Jewish businessman working for the Phoenix, a prosperous insurance company. So Mother had to leave, where to I do not know. Had we not been so blind we would have foreseen – and thousands of similar cases would have proved us right – what an excellent source of money the big, well-furnished flat could have proved for my mother had she lived longer. She died in the autumn of 1921 of cancer of the spine after what we believed had been a successful operation on her breast cancer in 1917.

I rarely remember dreams, and I seldom had nasty ones – except maybe in my early childhood. For a long time after my father's death, however, a nightmare kept recurring again and again: my father was still alive and I knew I had given away all his beautiful instruments and botanical books. What was he to do now that I had rashly and irretrievably destroyed the basis of his intellectual life? I am sure it was my guilty conscience that caused the dream, as I had cared so little for my parents between 1919 and 1921. This can be the only explanation, as I am not normally bothered with nightmares or a guilty conscience either.

My childhood and adolescence (1887–1910 or thereabouts) was mainly influenced by my father, not in the usual educational manner, but in a more ordinary way. This was due to his spending a lot more time at home than most men who work for a living and to my being at home, too. In my early years of learning I was taught by a private teacher who came to see me twice a week, and at grammar school we still had the blessed tradition of attending for twenty-five hours a week, mornings only. (On two afternoons only we had to attend for protestant religious education.)

I learnt a great deal on those occasions, although the result was not always related to the subject of religion. Time limitations concerning school commitments are a great asset. If a pupil feels inclined, he has time for thinking, and he can also take private lessons in the subjects which are not part of the curriculum. I can only find words of praise for my old school (Akademisches Gymnasium): I was rarely bored there, and when it did happen (our preparatory philosophical course was really bad), I would turn my attention to some other subject, my French translation, for example.

At this point I should like to add a remark of a more general kind. The discovery of chromosomes as the decisive factors in heredity seems to have given society the right to overlook other better-known but equally important factors such as communication, education and tradition. It is assumed that these were not so important because from the point of view of genetics they are not stable enough. This is quite true. However, there are cases such as that of Kaspar Hauser, for example, and that of a small group of Tasmanian 'Stone Age' children who were only recently brought to live in English surroundings and granted a first-class English upbringing, with the effect that they reached the educational level of upper-class Englishmen. Does this not prove to us that it takes both a code of chromosomes and civilized human surroundings to produce people of our kind? In other words, the intellectual level of every individual is bred by 'nature' and by 'nurture'. Schools are therefore (not as our Empress Maria Theresa liked to see it) invaluable for human guidance, and much less for political purposes. And a sound family background is just as important for preparing the soil for the seed the schools will sow. This is unfortunately a fact overlooked by those who claim that only the children of the less educated should attend schools for higher education (will their children be excluded for the same reasons?) and also by British High Society, where it is deemed upper class to replace family life by boarding school and considered a sign of nobility to leave home early. So even the present Queen had to part with her first-born and send him to such an institution. None of this is strictly speaking any of my concern. It only came to my mind when I once again realized how much I gained from the time I spent with my father as a young boy and how little I would have profited from school had he not been there. He actually knew far more than they had to offer, not because he had been forced to study it thirty years earlier, but because he was still interested. If I went into detail here, I should end up telling a long story.

Later on, when he had taken up botany and I had virtually devoured *The Origin of Species*, our discussions took on a

different character, certainly different from that conveyed at
school, where the theory of evolution was still banned from
our biology lessons and teachers of religious education were
advised to call it heresy. Of course I soon became an ardent
follower of Darwinism (and still am today, for that matter),
while Father, influenced by his friends, urged caution. The
link between natural selection and the survival of the fittest on
the one hand and Mendel's law and de Vries's theory of
mutation on the other had yet to be fully discovered. Even
today I don't know why zoologists have always tended to
swear by Darwin, while botanists appear to be rather more
reticent. However, one thing we all agreed on – and when I
say 'all', I particularly remember Hofrat Anton Handlisch,
who was a zoologist at the museum of natural history and the
one I knew and liked best of all my father's friends – we were
all unanimous in holding that the basis of evolutionary theory
was causal rather than finalistic; and that no special laws of
nature, such as vis viva, or an entelechy, or a force of
orthogenesis, etc., were at work in living organisms to abro-
gate or to counteract the universal laws of inanimate matter.
My religious teacher would not have been happy about this
view, but he did not concern me anyway.

Our family was accustomed to travelling in the summer.
This not only brightened my life, but also helped whet my
intellectual appetite. I remember one visit to England a year
before I started intermediate school (*Mittelschule*), when I
stayed with relatives of my mother at Ramsgate. The long,
wide beach was ideally suited for donkey rides and learning to
handle a bicycle. The strong tidal changes claimed my full
attention. Little bathing huts on wheels were set up along the
beach, and a man and his horse were always busy moving
these cabins up or down according to the tide. On the
Channel I first noticed that one could make out the funnel
smoke of distant boats on the horizon long before they
themselves appeared, a result of the curvature of the water-
surface.

In Leamington I met my great-grandmother at Madeira
Villa, and as she was called Russell and the street she lived in

was called 'Russell', I was convinced it was named after my late great-grandfather. An aunt of my mother's also lived there with her husband, Alfred Kirk, and six Angora cats. (In later years there were said to be twenty.) In addition she had an ordinary tomcat who would very often come home from his nocturnal adventures in a sad state, so he was given the name Thomas Becket (referring to the Archbishop of Canterbury who was killed in office by order of King Henry II) – not that this meant a great deal to me then, nor was it very appropriate.

It is thanks to my Aunt Minnie, Mother's youngest sister, who moved from Leamington to Vienna when I was five, that I learnt to speak fluent English long before I could write in German, let alone English. When I was finally introduced to the spelling and reading of the language I thought I knew so well, I was in for a surprise. It was thanks to my mother that half-days of English practice were launched. I was not too pleased about that at the time. We would walk from the Weiherburg down to the pretty and in those years still quiet little town of Innsbruck together, and Mother would say: 'Now we are going to speak English to each other the whole way – not another word of German.' And that is just what we did. I only realized later how much I profited from it to this day. Though forced to leave the country of my birth, I never felt a stranger abroad.

I seem to remember visiting Kenilworth and Warwick on our bicycle tours round Leamington. And on the way back to Innsbruck from England I remember seeing Bruges, Cologne, Coblenz – a steamboat took us up the Rhine – I remember Rüdesheim, Frankfurt, Munich, I think; then Innsbruck. I can recall the little boarding house which belonged to Richard Attlmayr.

From there I went to school for the first time, down to St Nikolaus, where I had private tuition, as my parents were afraid I had forgotten my ABC and my sums during the holiday and would fail my entrance exam in the autumn. In later years we nearly always went to the South Tyrol or Carinthia, and sometimes we would go to Venice for a few days in September. There is no end to the list of beautiful things I was given the chance to see in those days, things that no longer exist, due to the motor car, 'development' and new borders. I think few people then, let alone today, experienced such a happy childhood and adolescence as I did, even though I was an only child. Everyone was friendly towards me and we were all on good terms with each other. If only all teachers, including parents, would take to heart the necessity of mutual understanding! We cannot exert any lasting influence over those entrusted to us without it.

Maybe I ought to say something about my years at university between 1906 and 1910, as there might not be any chance of doing so later on. I mentioned earlier that Hasenöhrl and his carefully conceived four-year course (five hours a week!) influenced me more than anything else. Unfortunately I missed the last year (1910/11), as I could no longer postpone my national service. As it turned out this was not quite as unpleasant as I had anticipated, for I was sent to the beautiful old town of Cracow and I also spent a memorable summer near the Carinthian border (near Malborghet). Apart from Hasenöhrl's, I attended all the other mathematics lectures I could. Gustav Kohn gave his talks on projective geometry. His style, so severe and clear, left a lasting impression. Kohn would alternate from a pure synthetic method one year – without any formulas – to an analytical approach the next. There is in fact no better example for the existence of axiomatic systems. Through him duality in particular turned out to be a breathtaking phenomenon, differing somewhat in two- and three-dimensional geometry. He also proved to us the profound influence of Felix Klein's group theory on the development of mathematics. The fact that the existence of a fourth harmonic element has to be accepted as an axiom in a two-dimensional structure while it can easily be proved in a three-dimensional was to him the simplest illustration of Goedel's great theorem. There were so many things I learnt from Kohn which I would never have had the time to learn later on.

I attended Jerusalem's lectures on Spinoza – a memorable

experience for whoever listened to him. He talked about so many things, about Epicurus' ὁ ϑάνατος ουδέν προς ἡμάς ('Death is not man's enemy') and his ὀυδέν ϑαυμάζειν ('to wonder at nothing'), which Epicurus always kept in mind when philosophizing.

In my first year I also did qualitative chemical analysis, and certainly gained a lot from it. Skraup's lectures on inorganic chemical analysis were rather good; those on organic chemical analysis, which I read during the summer term, poor in comparison. They could have been ten times as good and still they would hardly have improved my understanding of nucleic acids, enzymes, antibodies and the like. As it was I could only feel my way ahead, led by intuition, which was none the less productive.

On 31 July 1914 my father turned up at my little office in the Boltzmanngasse to break the news that I had been called up. The Predilsattel in Carinthia was to be my first destination. We went off to buy two guns, a small one and a large one. Fortunately I was never forced to use them on either man or animal, and in 1938 during a search of my flat in Graz I handed them over to the good-natured official, just to be sure.

A few words about the war itself: my first posting, Predilsattel, was uneventful. Once, though, we had a false alarm. Our commanding officer, Captain Reindl, had arranged with confidants that in the event of Italian troops advancing up the wide valley towards the lake (Raiblersee), we were to be warned by smoke signals. It so happened that someone was baking potatoes or burning weeds just along the border. We were told to man the two watchposts and I was put in charge of the one on the left. We spent ten days up there before someone remembered to call us back down. Up there I learnt that springy floorboards (with only a sleeping-bag and blanket) are much more comfortable to sleep on than a solid floor. My other observation was of a different nature, something I never came across before or after. One night the guard on duty woke me up to report that he could see a number of lights moving up the slope opposite us, obviously heading toward our position. (Incidentally, this part of the mountain (Seekopf) had no paths at all.) I got out of my sleeping-bag and made my way through the connecting passage to the post to take a closer look. The guard was right about the lights, but they were St Elmo's fire on the top of our own wire abatis a couple of yards away, and the displacement against the background was only parallactic. This was because the observer himself was moving. When I stepped out of our spacious dug-out at night I would watch these pretty little fires on the tips of the grass that covered the roof. This was the only time I came across the phenomenon.

After spending much idle time there I was posted to Franzensfeste, then to Krems and then to Komorn. For a short time I had to serve at the front. I joined a small unit first at Gorizia, then at Duino. They were equipped with an odd naval gun. We eventually retired to Sistiana, and from there I was sent to a rather boring but none the less beautiful observation post near Prosecco, 900 feet above Trieste, where we had an even odder gun. My future wife Annemarie came to see me there, and on one occasion Prince Sixtus of Bourbon, the brother of the Empress Zita, visited our positions. He was not in uniform, and later I learnt that he was in fact our enemy as he was serving in the Belgian army. The reason for this was that the French did not allow any member of the Bourbon family to join their army. The aim of his visit at the time was to bring about a separate peace agreement between Austria-Hungary and the Entente Cordiale, which, of course, meant high treason against Germany. Unfortunately his plan never materialized.

My first encounter with Einstein's theory of 1916 was at Prosecco. I had so much time at my disposal, yet had great difficulties in understanding it. Nevertheless a number of marginal notes I made then still appear reasonably intelligent to me even now. As a rule Einstein would present a new theory in an unnecessarily complicated form, and never more so than in 1945, when he introduced the so-called 'asymmetric' unitary field theory. But perhaps that is not just characteristic of that great man, but nearly always happens when someone postulates a new idea. In the case of the above-mentioned theory Pauli told him there and then that it was unnecessary to introduce the complex quantities, because each of his tensor equations consisted of both a symmetric and a sheer symmetric part anyway. Only in 1952, in an article he wrote together with Mme B. Kaufman for a volume published to celebrate Louis de Broglie's sixtieth birthday, did he agree with my much simpler version by ingeniously excluding the so-called 'strong' version. This was a very important move indeed.

The last year or so of the war I spent as a 'meteorologist' first in Vienna, then Villach, then Wiener Neustadt and finally in Vienna again. This was a great asset to me, as I was spared the disastrous retreat of our badly torn front lines.

In March/April 1920 Annemarie and I got married. We moved soon after to Jena, where we took furnished lodgings. I was expected to add some up-to-date theoretical physics to Professor Auerbach's set lectures. We enjoyed the friendship and cordiality of both the Auerbachs, who were Jews, and of my boss Max Wien and his wife (they were anti-Semites by tradition, but bore no personal malice). Being on such good terms with them all was a great help to me. In 1933, the Auerbachs, I am told, saw no means of escape from the oppression and humiliation which Hitler's taking over (*Machtergreifung*) held in store for them but suicide. Eberhard Buchwald, a young physicist who had just lost his wife, and a couple called Eller with their two little sons were also amongst our friends in Jena. Mrs Eller came to see me here in Alpbach last summer (1959), a poor bereaved woman whose three men-folk had lost their lives fighting for a cause they did not believe in.

A chronological account of someone's life is one of the most boring things I can think of. Whether you are recalling incidents of your own life or that of someone else, you will rarely find more than the occasional experience or observation worth recounting – even if the historical order of events seems important to you at the time. That is why I am now going to give a short summary of the periods of my life, so that I can refer to them later without having to watch the chronological order.

The first period (1887–1920) ends with my marrying Annemarie and leaving Germany. I shall call it my first

Viennese Period. The second period (1920–7) I shall call 'My First Years of Roaming', as I was taken to Jena, Stuttgart, Breslau and finally to Zurich (in 1921). This period ends with my call to Berlin as Max Planck's successor. I had discovered wave mechanics during my stay in Arosa in 1925. My paper had been published in 1926. As a result of this I went on a two-month lecturing tour of North America, which prohibition had dried up successfully. The third period (1927–33) was a rather nice one. I shall call it 'My Teaching and Learning'. It ended with Hitler's assumption of power, the so-called *Machtergreifung*, in 1933. While completing the summer term of that year I was already busy sending my belongings to Switzerland. At the end of July I left Berlin to spend my holidays in the South Tyrol. The South Tyrol had become Italian under the Treaty of St Germain, so it was still accessible to us with our German passports, whereas Austria was not. Prinz Bismarck's great successor had succeeded in imposing a blockade in Austria which became known as the *Tausendmarksperre*. (My wife, for instance, could not visit her mother on her seventieth birthday. His Excellency's authorities did not give her permission). I did not go back to Berlin after the summer, but instead handed in my resignation, which remained unanswered for a long time. In fact they then denied ever having received it, and when they learnt I had been awarded the Nobel Prize for physics, they flatly refused to accept it.

The fourth period (1933–9) I shall call 'My Later Years of Roaming'. As early as spring 1933 F. A. Lindemann (later Lord Cherwell) offered me a 'living' in Oxford. This was on the occasion of his first visit to Berlin, when I happened to mention my distaste for the present situation. He faithfully kept his word. And so my wife and I took to the road in a little BMW acquired for the occasion. We left Malcesine and via Bergamo, Lecco, St Gotthard, Zurich and then Paris we reached Brussels, where a Solvay Congress was being held. From there we went to Oxford; we did not travel together. Lindemann had already taken the necessary steps to make me a fellow of Magdalen College, though I received the greater part of my pay from ICI.

When, in 1936, I was offered a chair at Edinburgh University and another at Graz, I chose the latter, an extremely foolish thing to do. Both the choice and the outcome were unexampled, though the outcome was a lucky one. Of course I was more or less undermined by the Nazis in 1938, but by then I had already accepted a call to Dublin, where de Valera was about to found the Institute for Advanced Studies. Loyalty towards his own university would never have allowed Edinburgh's E.T. Whittaker, de Valera's former teacher, to suggest me for the post had I gone to Edinburgh in 1936. As it was, Max Born was appointed in my stead. Dublin proved a hundred times better for me. Not only would the work in Edinburgh have been a great burden to me, but so would the position of enemy alien in Great Britain throughout the war.

Our second 'escape' took us from Graz, via Rome, Geneva and Zurich to Oxford where our dear friends, the Whiteheads, put us up for two months. This time we had to leave our good little BMW, 'Grauling', behind, as it would have been too slow, and besides, I no longer possessed a driving licence. The Dublin Institute was not yet 'ready', and so my wife, Hilde, Ruth and I went to Belgium in December 1938. First I held lectures (in German!) at the University of Ghent as guest professor; this was for the 'fondation Franqui-Seminar'. Later on we spent about four months in Lapanne by the sea. It was a lovely time – despite the jellyfish. It was also the only time I ever came across the phosphorescence of the sea. In September 1939, the first month of the Second World War, we left for Dublin via England. With our German passports we were still enemy aliens to the British, but obviously thanks to de Valera's letters of reference we were granted transit. Perhaps Lindemann pulled a few strings on that occasion too, despite the rather unpleasant encounter we had had a year before. He was after all a very decent man, and I am convinced that as his friend Winston's advisor in matters of physics he proved invaluable in the defence of Britain during the war.

The fifth period (1939–56) I shall call 'My Long Exile', but without the bitter associations of the word, as it was a wonderful time. I would never have got to know this remote and beautiful island otherwise. Nowhere else could we have lived through the Nazi war so untouched by problems that it is almost shameful. I can't imagine spending seventeen years in Graz 'treading water', with or without the Nazis, with or without the war. Sometimes we would quietly say amongst ourselves: 'Wir danken's unserem Führer' ('We owe it to our Führer').

The sixth period (1956–?) I shall call 'My Late Viennese Period'. As early as 1946 I had been offered an Austrian chair again. When I told de Valera about it he urgently advised me against it, pointing to the unsettled political situation in Central Europe. He was quite right in that respect. But while he was so kindly disposed towards me in many ways, he showed no concern for my wife's future should anything happen to me. All he could say was that he wasn't sure what would happen to his wife in such a situation either. So I told them in Vienna that I was keen on going back, but that I wanted to wait for matters to return to normal. I told them that because of the Nazis I had been forced to interrupt my work twice already and start all over again elsewhere; a third time would certainly put an end to it altogether.

Looking back, I can see that my decision was right. Poor Austria had been raped and was a sad place to live in those days. My petition addressed to the Austrian authorities for a pension for my wife as a kind of reparation was in vain in spite of the fact that they seemed keen to make amends. The poverty was too great then (and still is today in 1960, for that matter) to make allowances for certain individuals and deny them to almost all others. Thus I spent ten more years in Dublin, which turned out to be of great value to me. I wrote quite a number of short books in English (published by Cambridge University Press) and continued my studies on the 'asymmetric' general theory of gravitation, which appears to be disappointing. And last but not least there were the two successful operations in 1948 and 1949 by Mr Werner, who removed the cataracts from both my eyes. When the time had come, Austria very generously restored me to my former position. I also received a new appointment to Vienna

University (extra status), although at my age I could only expect two and a half years in office. I owe all this mainly to my friend Hans Thirring, and to the Minister of Education, Dr Drimmel. At the same time my colleague Robracher successfully pushed the new law for the status of Professor Emeritus and thus also supported my cause.

This is where my chronological summary ends. I hope to add a few ideas or details here and there that are not too boring. I must refrain from drawing a complete picture of my life, as I am not good at telling stories; besides, I would have to leave out a very substantial part of this portrait, i.e. that dealing with my relationships with women. First of all it would no doubt kindle gossip, secondly it is hardly interesting enough for others, and last but not least I don't believe anyone can or may be truthful enough in those matters.

This summing-up was written early this year. It now gives me pleasure to read through it occasionally. But I have decided not to continue – there would be no point.

E.S.   November 1960